The Islamic Threat

THE
ISLAMIC THREAT
Myth or Reality?

JOHN L. ESPOSITO

New York   Oxford
OXFORD UNIVERSITY PRESS
1992

Oxford University Press

Oxford   New York   Toronto
Delhi   Bombay   Calcutta   Madras   Karachi
Kuala Lumpur   Singapore   Hong Kong   Tokyo
Nairobi   Dar es Salaam   Cape Town
Melbourne   Auckland

and associated companies in
Berlin   Ibadan

Copyright © 1992 by John L. Esposito

Published by Oxford University Press, Inc.
200 Madison Avenue, New York, New York 10016

Oxford is a registered trademark of Oxford University Press, Inc.

Library of Congress Cataloging-in-Publication Data
Esposito, John L.
The Islamic threat : myth or reality? /
John L. Esposito.
p.   cm.   Includes bibliographical references and index.
ISBN 0-19-507184-0
1. Islam—20th century.   2. Islamic countries—Relations—Europe.
3. Europe—Relations—Islamic countries.   I. Title.
BP60.E84   1992   297′.99′04—dc20   92-8959

2 4 6 8 9 7 5 3

Printed in the United States of America
on acid free paper

For my parents,
John and Mary Esposito

Preface

This is a book I never intended to write. However, over the years I have become increasingly aware that a series of world events—particularly those stretching from the Ayatollah Khomeini and the Iranian revolution to Saddam Hussein and the Gulf War—have reinforced mutually destructive stereotypes in the Muslim world and in the West. The underlying presuppositions in many of the questions put to me at conferences and speaking engagements across the country and in various media exemplified this problem. Whether in university lectures, community talks, or government briefings on Islam or on Muslim politics, images of a militant Islam would surface, of a faith that is regarded as particularly prone to religious extremism, fanaticism, and warfare. For many, the Ayatollah Khomeini and Saddam Hussein epitomized a decade of radical Islamic fundamentalism, militant and anti-Western, seen as part of a centuries-long historical pattern. This popular attitude was also reflected in articles by prominent political commentators. As I looked toward the postwar period and the inevitable issue of the creation of a New World Order as well as the growing populist calls within the Muslim world for greater political liberalization and democratization, I felt a pressing need to address the issue of an Islamic threat and to place it in historical perspective.

Although I have published many books, this one has been both the easiest and most difficult to write: easy because it grows out of and builds upon previous studies of contemporary Islam in general and Islamic revivalism or fundamentalism in particular; difficult because the mere mention of the topic conjures up strongly held convictions, deep resentments, and fears. Like it or not, the history of relations between Muslims and Christians, between the Muslim world and the West, both in the past and the present, has been marked by episodes of confrontation and conflict. This book will have its admirers and detractors, with few readers not taking sides.

For many in the West for whom Islam self-evidently refers to an aggressive, hostile, and anti-American religious tradition and people, this volume will appear too uncritical, seemingly unaware of the "Sword of Islam." For others, who dismiss all negative judgments about Muslim activities as simply the result of neoimperialism and anti-Islamic sentiment, the highlighting of authoritarianism and violence in the name of Islam will be disturbing.

The reader should be forewarned that a certain selectivity of coverage and a degree of generalization are unavoidable. Since the purpose of this book is to address the question of whether there is an Islamic threat to the West and to explore related questions—the sources of confrontation and conflict; the nature of Islamic fundamentalism; the reasons for the perception of Islam as a threat to the West and of the West as a threat to the Muslim world, and its impact and implications for U.S. policy—this study will necessarily look at those aspects of Islamic religion, history, and politics that are directly relevant. Thus in discussing the historical record of confrontation and conflict between Islam and Christianity, or between the Muslim world and the West, I will rarely focus on the positive history of Muslim–Christian relations. At many points generalizations and stereotypes will be employed simply because they are part of the history and perceptions of two world civilizations. Sadly, both in the past and the present, religious and secular biases and historical experiences have produced a set of stereotypes that resurface in times of conflict, Crusader versus jihadist, or the modern, developed, secular, rational Western mind versus the traditional, underdeveloped, religious, emotional, irrational Muslim mind.

I will argue throughout this book that we must move beyond a monolithic worldview that sees Muslims and the Muslim world (both governments and social movements) as a unity. Similarly, Europe or the West should also be appreciated in their diversity and complexity. Yet since both sides tend to consider and talk about each other as a bloc, the language of "Islam" and the "West" will of necessity be employed at times.

A number of colleagues have been generous with their time in reviewing earlier drafts of this book. James P. Piscatori, Yvonne Y. Haddad, John O. Voll, and Tamara Sonn read the entire manuscript. John Entelis and David Long commented on chapter 6. Although they bear no responsibility for my opinions and judgments, all played a significant role in offering advice and encouragement. The College of the Holy Cross once

again provided a much-needed year-long research grant. Cynthia A.
Read is the kind of editor every author hopes for. She and her colleagues
at Oxford University Press embody the highest standards of excellence.
My wife, Jean, has been both a constant source of encouragement
throughout the years and my most demanding critic.

In the coming decade Muslim politics will remain fragile and combustible, as will Muslim relations with the West. The Muslim world continues
to expand beyond the more than forty-five Muslim nations. In addition,
Muslim populations continue to grow in the former Soviet Union,
China, India, Europe, and the United States. In an increasingly interdependent world, it becomes even more imperative to deal with political
realities dispassionately and constructively and to transcend stereotypes.

Wayland, Mass.   J.L.E.
January 1992

Contents

Introduction, 3

1. Contemporary Islam: Reformation or Revolution? 7

2. Islam and the West: Roots of Conflict, Cooperation, and Confrontation, 25

3. The West Triumphant: Muslim Responses, 47

4. Islam and the State: Dynamics of the Resurgence, 77

5. Islamic Organizations: Soldiers of God, 119

6. "Islamic Fundamentalism" and the West, 168

Notes, 213
Select Bibliography, 226
Index, 231

The Islamic Threat

Introduction

There are a good many people who think that the war between
communism and the West is about to be replaced by a war between
the West and Muslims.

WILLIAM PFAFF, "Help Algeria's Fundamentalists,"
The New Yorker, January 28, 1991

Are Islam and the West on an inevitable collision course? Are Islamic
fundamentalists medieval fanatics? Are Islam and democracy incompatible? Is Islamic fundamentalism a threat to stability in the Muslim world
and to American interests in the region? These are critical questions for
our times that come from a history of mutual distrust and condemnation.

From the Ayatollah Khomeini to Saddam Hussein, for more than a
decade the vision of Islamic fundamentalism or militant Islam as a threat
to the West has gripped the imaginations of Western governments and
the media. Khomeini's denunciation of America as the "Great Satan,"
chants of "Death to America," the condemnation of Salman Rushdie
and his *Satanic Verses,* and Saddam Hussein's call for a jihad against
foreign infidels have reinforced images of Islam as a militant, expansionist religion, rabidly anti-American and intent upon war with the West.

Despite many common theological roots and beliefs, throughout history Muslim–Christian relations have often been overshadowed by conflict as the armies and missionaries of Islam and Christendom have
struggled for power and for souls. This confrontation has involved such
events as the defeat of the early Byzantine (eastern Roman) empire by
Islam in the seventh century; the fierce battles and polemics of the
Crusades during the eleventh and twelfth centuries; the expulsion of the
Moors from Spain and the Inquisition; the Ottoman threat to Europe;
European (Christian) colonial expansion and domination in the eighteenth and nineteenth centuries; the political and cultural challenge of

the superpowers (America and the Soviet Union) in the latter half of the twentieth century; the creation of the state of Israel; the competition of Christian and Muslim missionaries for converts in Africa today; and the contemporary reassertion of Islam in politics.

"Islamic fundamentalism" has often been regarded as a major threat to the regional stability of the Middle East and to Western interests in the broader Muslim world.[1] The Iranian Revolution, attacks on Western embassies, hijackings and hostage taking, and violent acts by groups with names like the Army of God (Jund Allah), Holy War (al-Jihad), the Party of God (Hizbullah), and Salvation from Hell have all signaled a militant Islam on a collision course with the West. Uprisings in the Muslim republics of the Soviet Union, in Kosovo in Yugoslavia, in Indian Kashmir, in Sinkiang in China, and on the West Bank and in Gaza, and more recently, Saddam Hussein's attempted annexation of Kuwait, have reinforced images of an expansive and potentially explosive Islam in global politics.

With the triumph of the democratization movement in Eastern Europe and the breakup of the Soviet empire, Islam constitutes the most pervasive and powerful transnational force in the world, with one billion adherents spread out across the globe. Muslims are a majority in some forty-five countries ranging from Africa to Southeast Asia, and they exist in growing and significant numbers in the United States, the Soviet Union, and Europe. For a Western world long accustomed to a global vision and foreign policy predicated upon superpower rivalry for global influence if not dominance—a U.S.–Soviet conflict often portrayed as a struggle between good and evil, capitalism and communism—it is all too tempting to identify another global ideological menace to fill the "threat vacuum" created by the demise of communism.

However diverse in reality, the existence of Islam as a worldwide religion and ideological force embracing one fifth of the world's population, and its continued vitality and power in a Muslim world stretching from Africa to Southeast Asia, will continue to raise the specter of an Islamic threat. In the 1990s it is important that the vacuum created by the end of the Cold War not be filled by exaggerated fears of Islam as a resurgent "evil empire" at war with the New World Order and a challenge to global stability: "No matter how and when the war ends, Islamic rage already threatens the stability of traditionally pro-Western regimes from Morocco to Jordan to Pakistan."[2] Belief that a clash of worldviews, values, and civilizations is leading to an impending confrontation between Islam and the West is reflected in headlines and articles with ominous titles like these: "Still Fighting the Crusades," "The New Crescent in Crisis: The Global Intifada," "Rising Islam May Overwhelm the

West," "The Roots of Muslim Rage," "The Islamic War Against Modernity," and the "Arc of Crisis." While such phrases capture public attention and the popular imagination, they exaggerate and distort the nature of Islam, the political realities of the Muslim world and its diverse relations with the West. They also reinforce an astonishing degree of ignorance and cultural stereotyping of Arabs and Islam. For many in the West it is axiomatic that Arabs are nomads or oil shaykhs, denizens of the desert and harems, an emotional, combative, and irrational people. Islam is often equated with holy war and hatred, fanaticism and violence, intolerance and the oppression of women.

As Western leaders attempt to forge the New World Order, transnational Islam may increasingly come to be regarded as the new global monolithic enemy of the West: "To some Americans, searching for a new enemy against whom to test our mettle and power, after the death of communism, Islam is the preferred antagonist. But, to declare Islam an enemy of the United States is to declare a second Cold War that is unlikely to end in the same resounding victory as the first."[3] Fear of the Green Menace (green being the color of Islam) may well replace that of the Red Menace of world communism.

Islam and Islamic movements constitute a religious and ideological alternative or challenge and in some instances a potential danger to Christianity and the West. However, distinguishing between a religious or ideological alternative or challenge and a direct political threat requires walking the fine line between myth and reality, between the unity of Islam and the diversity of its multiple and complex manifestations in the world today, between the violent actions of the few and the legitimate aspirations and policies of the many. Unfortunately, American policymakers, like the media, have too often proved surprisingly myopic, viewing the Muslim world and Islamic movements as a monolith and seeing them solely in terms of extremism and terrorism. While this is understandable in light of events in Iran and Lebanon and the Gulf crisis of 1990–91, it fails to do justice to the complex realities of the Muslim world and can undermine relations between the West and Islam.

The Islamic Threat: Myth or Reality? will place the challenge or threat of Islam in perspective, and discuss the vitality of Islam as a global force and the history of its relations with the West. Case studies of Muslim countries and Islamic movements will demonstrate the diversity in geography, politics, ideological and organizational orientation, tactics, and foreign policy of the Islamic resurgence.

Finally, the issues facing Islam and the West in the nineties and the potential challenge or threat of Islam to the West will be explored in light of the Salman Rushdie affair, the Gulf War of 1990–91, the New

World Order, and the challenge of democratization in the Muslim world. Do the West and the Muslim world share common interests and values or does a clash of worldviews loom on the horizon? Is the drive for liberalization and democracy in the Muslim world by Islamic movements merely a tactical means to an end? What would the implications of Islamic societies be with regard to pluralism, human rights, and the status of women and minorities? How can U.S. policy transform the specter of an Islamic threat into the pursuit of a common global future?

The Economist put these issues in perspective in 1990 in an imaginary dialogue between a Christian and a Muslim religious leader who spoke as follows:

> It distresses me that so many people seem to think the next period of history will be a fight between your part of the world and mine. It is true that we live elbow to elbow with each other . . . it is also true that our elbows have banged painfully together many times in the past. But almost 2,000 years after the birth of your Jesus, and more than 1,400 years after the birth of our Muhammad, let me start by asking whether it really has to happen all over again.[4]

1

Contemporary Islam: Reformation or Revolution?

Much of the reassertion of religion in politics and society has been subsumed under the term *Islamic fundamentalism*. Although "fundamentalism" is a common designation, in the press and increasingly among academics it is used in a variety of ways. For a number of reasons it tells us everything and yet, at the same time, nothing. First, all those who call for a return to foundational beliefs or the fundamentals of a religion may be called fundamentalist. In a strict sense this could include all practicing Muslims, who accept the Quran as the literal word of God and the Sunnah (example) of the Prophet Muhammad as a normative model for living.

Second, our understanding and perceptions of fundamentalism are heavily influenced by American Protestantism. *Webster's Ninth New Collegiate Dictionary* defines the term *fundamentalism*—as "a movement in 20th century Protestantism emphasizing the literally interpreted Bible as fundamental to Christian life and teaching." For many liberal or mainline Christians, "fundamentalist" is pejorative or derogatory, being applied rather indiscriminately to all those who advocate a literalist biblical position and thus are regarded as static, retrogressive, and extremist. As a result, fundamentalism often has been regarded popularly as referring to those who are literalists and wish to return to and replicate the past. In fact, few individuals or organizations in the Middle East fit such a stereotype. Indeed, many fundamentalist leaders have had the best education, enjoy responsible positions in society, and are adept at harnessing the latest technology to propagate their views and create viable modern institutions such as schools, hospitals, and social service agencies.

Third, "fundamentalism" is often equated with political activism, extremism, fanaticism, terrorism, and anti-Americanism. Yet while some engage in radical religiopolitics, most, as we shall see, work within the established order.

Perhaps the best way to appreciate the facile use of "fundamentalism" and its inadequacy (the many faces and postures of fundamentalism) is to consider the following. This term has been applied to the governments of Libya, Saudi Arabia, Pakistan, and Iran. Yet what does that really tell us about these states other than the fact that their rulers have appealed to Islam to legitimate their rule or policies? Muammar Qaddafi has claimed the right to interpret Islam, questioned the authenticity of the traditions of the Prophet Muhammad, silenced the religious establishment as well as the Muslim Brotherhood, and advocated a populist state of the masses. The rulers of Saudi Arabia, by contrast, have aligned themselves with the *ulama* (clergy), preached a more literalist and rigorous brand of Islam, and used religion to legitimate a conservative monarchy. Qaddafi's image as an unpredictable, independent supporter of worldwide terrorism stands in sharp relief beside the low-key, conservative, pro-American King Fahd. Similarly, contrast the foreign policy of the clerically run Shii state of Iran with the military, lay regime (1977–88) which implemented Pakistan's Islamic system (*nizam-i-Islam*) under Gen. Zia ul-Haq. Iran under the Ayatollah Khomeini was highly critical—even condemnatory—of the West, often at odds with the international community, and regarded as a radical terrorist state, while Pakistan under the Islamically oriented Zia ul-Haq was a close ally of the United States who enjoyed warm relations with the West and the international community and was generally regarded as moderate.

I regard "fundamentalism" as too laden with Christian presuppositions and Western stereotypes, as well as implying a monolithic threat that does not exist; more fitting general terms are "Islamic revivalism" or "Islamic activism," which are less value-laden and have roots within the Islamic tradition. Islam possesses a long tradition of revival (*tajdid*) and reform (*islah*) which includes notions of political and social activism dating from the early Islamic centuries to the present day.[1] Thus I prefer to speak of Islamic revivalism and Islamic activism rather than of Islamic fundamentalism.

The Resurrection of Religion: Modernization and Development Theory Confounded

Just as European imperialism and colonialism had been based upon and legitimated by a modernity that was Western in its origins and forms, so too the development of non-Western regions including the Muslim world was based upon a theory of modernization that equated development with the progressive Westernization and secularization of society. In particular, secularization was regarded as a sine qua non for modernization: "Political development includes, as one of its basic processes, the secularization of polities, the progressive exclusion of religion from the political system."[2] Both the indigenous elites, who guided government development programs in newly emerging Muslim states, and their foreign patrons and advisers were Western-oriented and Western-educated. All proceeded from a premise that equated modernization with Westernization. The clear goal and presupposition of development was that every day and in every way things should become more modern (i.e., Western and secular), from cities, buildings, bureaucracies, companies, and schools to politics and culture. While some warned of the need to be selective, the desired direction and pace of change were unmistakable. Even those Muslims who spoke of selective change did so within a context which called for the separation of religion from public life. Western analysts and Muslim experts alike tended to regard a Western-based process of modernization as necessary and inevitable and believed equally that religion was a major hindrance to political and social change in the Muslim world.

The conceptual world which resulted was one of clear dichotomies: tradition versus change, fundamentalism versus modernism, stagnation versus progress and development. Science and technology were seen as powerful aids in this process of secular development. However, the reality proved to be far different from the ideal. Modernization as Westernization and secularization remained primarily the preserve of a small minority elite of society. Most important, the secularization of processes and institutions did not easily translate into the secularization of minds and culture. While a minority accepted and implemented a Western secular worldview, the majority of most Muslim populations did not internalize a secular outlook and values. Daniel Crecelius's perceptive observation regarding Egypt, long regarded as among the more modern Muslim societies, ran counter to the prevailing wisdom but proved true for most Muslim countries:

> Most studies on the process of modernization or secularism recognize the necessity for all systems by which man lives, the psychological and intellectual, no less than the political and economic, to undergo transformation. We do not find this change in Egypt, whether at the level of the state or society, except among a small minority of Westernized individuals. Traditional beliefs, practices, and values reign supreme among Egypt's teeming village population and among the majority of its urban masses. It should

be emphasized that adherence to tradition is not confined to any single class or group of occupations, but is characteristic of a broad spectrum of all Egyptian social classes."[3]

For many, the contemporary revival of Islam challenged the received wisdom and seemed to deal a death blow to reason and common sense. The most forceful manifestations of the Islamic resurgence have occurred in the more advanced and "modernized" (seemingly secular) countries of the Muslim world such as Egypt, Iran, Lebanon, and Tunisia. In a very real sense, Islamic revivalism has often been seen and experienced as a direct threat to the ideas, beliefs, practices, and interests of Muslim secular elites as well as Western governments and multinational corporations. This clash of worldviews has reinforced the Western tendency to see Islamic activism as extremism and fanaticism, as an antimodern return to the past rather than the projection of an alternative vision for society. Because it does not conform to modern secular presuppositions, to the West's most cherished beliefs and values, Islamic activism is regarded as a dangerous, irrational and countercultural movement.

Yet Islamic organizations have attracted the educated and professionals (teachers, engineers, lawyers, scientists, bureaucrats, the military). Many of the leaders of Islamic organizations are graduates of major universities from faculties of medicine, science, and engineering. Modern technology has been harnassed by conservative clergy and political activists alike to organize and mobilize mass support as well as disseminate their message of religion and sociopolitical activism. The widespread use of radio, television, audio and videocassettes, computers, and fax machines has made for a more effective communication of Islam nationally and transnationally. Thus technology and communications have purveyed not simply a modern secular culture but also a revitalized and, at times, transnational Islam. Religious leaders, who were initially resistant, have come to depend upon modern technology. Village Muslims no longer live in relatively isolated worlds solely dependent upon their local religious leaders for knowledge of Islam. Television, radio, and audiocassettes now expose them to a diversity of voices (preachers) and messages or interpretations of Islam. Just as Christian televangelists can now preach their messages not just to those in their parishes but to audiences from rural Oregon to New York City and to peoples throughout the world, so too modern science and technology have been harnessed to preach the message of Islam. Whereas in the past people were limited by the realities of time and space and of language and government censorship, today Muslims throughout the world have available to them inexpensive translations and recordings of Islamic materials. They can be broadcast legally or easily shipped across borders, as witnessed by the spread of the writings and sermons of Khomeini throughout Iran during the Shah's rule, or the availability of the writings and cassettes of popular preachers and activists throughout the world. Thus modernization has not simply led to progressive secularization but instead has been a major factor in the resurgence of Islam in Muslim societies.

The Islamic Resurgence

Islam reemerged as a potent global force in Muslim politics during the 1970s and 1980s.[4] The scope of the Islamic resurgence has been worldwide, embracing much of the Muslim world from the Sudan to Indonesia. Heads of Muslim governments as well as opposition groups increasingly appealed to religion for legitimacy and to mobilize popular support. Islamic activists have held cabinet-level positions in Jordan, the Sudan, Iran, Malaysia, and Pakistan. Islamic organizations constitute the leading opposition parties and organizations in Egypt, Tunisia, Algeria, Morocco, the West Bank and Gaza, and Indonesia. Where permitted, they have participated in elections and served in parliament and in city government. Islam has been a significant ingredient in nationalist struggles and resistance movements in Afghanistan, the Muslim republics of the former Soviet Central Asia, and Kashmir, and in the communal politics of Lebanon, India, Thailand, China, and the Philippines.

Islamically oriented governments have been counted among America's staunchest allies (Saudi Arabia and Pakistan) and most vitriolic enemies (Libya and Iran). Islamic activist organizations have run the spectrum from those who work within the system—such as the Muslim Brotherhoods in Egypt, Jordan, and the Sudan—to radical revolutionaries like Egypt's Society of Muslims (known more popularly as Takfir wal-Hijra, Excommunication and Flight) and al-Jihad (Holy War), or Lebanon's Hizbullah (Party of God) and Islamic Jihad, which have resorted to violence in their attempts to overthrow prevailing political systems.

Yet to speak of a contemporary Islamic revival can be deceptive, if this implies that Islam had somehow disappeared or been absent from the Muslim world. It is more correct to view Islamic revivalism as having led to a higher profile of Islam in Muslim politics and society. Thus what had previously seemed to be an increasingly marginalized force in Muslim public life reemerged in the seventies—often dramatically—as a vibrant sociopolitical reality. Islam's resurgence in Muslim politics reflected a growing religious revivalism in both personal and public life that would sweep across much of the Muslim world and have a substantial impact on the West in world politics.

The indices of an Islamic reawakening in personal life are many: increased attention to religious observances (mosque attendance, prayer, fasting), proliferation of religious programming and publications, more emphasis upon Islamic dress and values, the revitalization of Sufism (mysticism). This broader-based renewal has also been accompanied by Islam's reassertion in public life: an increase in Islamically oriented governments, organizations, laws, banks, social welfare services, and educational institutions. Both governments and opposition movements have turned to Islam to enhance their authority and muster popular support. Governmental use of Islam has been illustrated by a great spectrum of leaders in the Middle East and Asia: Libya's Muammar Qaddafi, Sudan's Gaafar Muhammad Nimeiri, Egypt's Anwar Sadat, Iran's Ayatollah Khomeini, Pakistan's Zia ul-Haq, Bangladesh's Muhammad Ershad, Malaysia's Muhammad Mahathir. Most rulers and governments, including more secular states such as Turkey and Tunisia, becoming aware of the potential strength of Islam, have shown increased sensitivity to and anxiety about Islamic issues. The Iranian Revolution of 1978–79 focused attention on "Islamic fundamentalism" and with it the spread and vitality of political Islam in other parts of the Muslim world. However, the contemporary revival has its origins and roots in the late sixties and early seventies, when events in such disparate areas as Egypt and Libya as well as Pakistan and Malaysia contributed to experiences of crisis and failure, as well as power and success, which served as catalysts for a more visible reassertion of Islam in both public and private life.

The Experience of Failure and the Quest for Identity

Several conflicts (e.g., the 1967 Arab–Israeli war, Chinese–Malay riots in Malaysia in 1969, the Pakistan–Bangladesh civil war of 1971, and the Lebanese civil war of the midseventies) illustrate the breadth and diversity of these turning points or catalysts for change. For many in the Arab and broader Muslim world, 1967 proved to be a year of catastrophe as well as a historic turning point. Israel's quick and decisive defeat of Arab forces in what was remembered as the Six-Day War, the Israeli capture and occupation of the Golan Heights, Sinai, Gaza, the West Bank, and East Jerusalem, constituted a devastating blow to Arab/Muslim pride, identity, and self-esteem. Most important, the loss of Jerusalem, the third holiest city of Islam, assured that Palestine and the liberation of Jerusalem would not be regarded as a regional (Arab) issue but rather as an Islamic cause throughout the Muslim world. The defense of Israel is dear to many Jews throughout the world. Likewise, for Muslims who retain a sense of membership in a transnational community of believers (the *ummah*), Palestine and the liberation of Jerusalem are strongly seen as issues of Islamic solidarity. As anyone who works in the Muslim world can attest, Israeli control of the West Bank, Gaza, and Jerusalem as well as U.S.–Israeli relations are topics of concern and bitter debate among Muslims from Nigeria and the Sudan to Pakistan and Malaysia, as well as among the Muslims of Europe and the United States.

The aftermath of the 1967 war, remembered in Arab literature as the "disaster," witnessed a sense of disillusionment and soul-searching that gripped both Western-oriented secular elites as well as the more Islamically committed, striking at their sense of pride, identity, and history.[5] Where had they gone wrong? Both the secular and the Islamically oriented sectors of society now questioned the effectiveness of nationalist ideologies, Western models of development, and Western allies who had persisted in supporting Israel. Despite several decades of independence and modernization, Arab forces (consisting of the combined military might of Egypt, Jordan, and Syria) had proved impotent. A common critique of the military, political, and sociocultural failures of Western-oriented development and a quest for a more authentic society and culture emerged—an Arab identity less dependent upon the West and rooted more indigenously in an Arab/Islamic heritage and values. Examples from Malaysia, Pakistan, and Lebanon reflect the turmoil and soul-searching that occurred in many parts of the Muslim world.

Because Islam is often equated with the Middle East, we tend to forget that the largest Muslim populations are to be found in Asia (Indonesia, Pakistan, Bangladesh, and India). Asia too proved to be a major theater for the growth of Islamic revivalism. In Southeast Asia, Chinese–Malay communal riots in Kuala Lumpur in 1969 signaled growing Malay Muslim discontent and alienation. Though dominant politically, Malay Muslims lagged behind the more prosperous urban-based Chinese (non-Muslim) minority. In response to charges that the Chinese enjoyed disproportionate economic and educational advantages, a perceived threat to Malay status and identity, the Malaysian government responded by implementing an affirmative action–like plan (*bhumiputra,* sons of the soil) of incentives and quotas to strengthen Malay Muslim life. Greater emphasis on Malay identity, language, values, and community contributed to the attraction and growth of Islamic revivalism in a culture where many regard it as axiomatic that to be Malay is to be Muslim.

The Pakistan–Bangladesh civil war in 1971 changed the map of South Asia when the Islamic Republic of Pakistan, established in 1947 as a Muslim homeland, lost its eastern section. The loss of East Pakistan and its re-creation as Bangladesh raised serious questions about the nature (or perhaps better, the failure) of Pakistan's Islamic identity and ideology. At the same time Pres. Zulfikar Ali Bhutto, a secular socialist, for economic and strategic reasons increasingly appealed to Islam in the seventies to establish Pakistan's ties with its oil-rich Arab/Muslim brothers in the Gulf states, who offered economic aid, and revenue in the form of wages of Pakistani military and laborers hired to work in the Gulf. This tilt toward the Gulf unleashed a process in which Islam moved from the periphery to center stage as both the government and the opposition used it to legitimate their competing claims and to gain popular support.

In Lebanon the Shii Muslims, long a minority in a Christian-dominated system, increasingly called for greater political representation and socioeconomic reforms to better reflect demographic changes which had resulted in a Muslim majority. A charismatic religious leader, Imam Musa Sadr, appealed to Shii identity, history, and symbols to organize and mobilize members of the Shii community into what in the midseventies would become the Movement for the Dispossessed, more commonly known today as AMAL. In the aftermath of the Iranian revolution, Lebanon would see the rise of more radical Islamic groups such as Hizbullah (Party of God), equally concerned with issues of identity and power.

The forms that the Islamic revival has taken have varied almost infinitely from one country to another. However, there are recurrent themes: a sense that existing political, economic, and social systems had failed; a disenchantment with, and at times a rejection of, the West; a quest for identity and greater authenticity; and the conviction that Islam provides a self-sufficient ideology for state and society, a valid alternative to secular nationalism, socialism, and capitalism.

The experience of failure triggered an identity crisis which led many to question the direction of political and social development and to turn inward for strength and guidance. The Western-oriented policies of governments and elites appeared to have failed. The soul-searching and critique of the sociopolitical realities of the Arab and Muslim world, which followed the 1967 war and the crises in Pakistan, Malaysia, and Lebanon, extended to other Muslim areas, embraced a broad spectrum of society, and raised many questions about the direction and accomplishments of development. More often than not, despite the hopes aroused by independence, the mixed record of several decades of existence was a challenge to the legitimacy and effectiveness of modern Muslim states. A crisis mentality fostered by specific events and the general impact and disruption of modernity spawned a growing disillusionment, a sense of failure.

Politically, modern secular nationalism was found wanting. Neither liberal nationalism nor Arab nationalism/socialism had fulfilled its promises. Muslim governments seemed less interested and successful in establishing their political legitimacy and creating an ideology for national unity than in perpetuating autocratic rule. The Muslim world was still dominated by monarchs and military or ex-military rulers, political parties were banned or restricted, and elections were often rigged. Parliamentary systems of government and political parties existed at the sufferance of rulers whose legitimacy, like their security, depended on a loyal military and secret police. Many were propped up by and dependent upon foreign governments and multinational corporations as well.

Charges of corruption and of concentration and maldistribution of wealth found a ready reception as one looked at individual countries and the region. The disparity between rich and poor was striking in urban areas, where the neighborhoods and new suburbs of the wealthy few stood in stark contrast to the deteriorating dwellings and sprawling shantytowns of the many. The vast chasm between rich and poor was even more pronounced between Arab oil states, which ironically tended to be among the least populated countries (Kuwait, Saudi Arabia, the Emirates) and the greater number of poor, densely populated countries (Egypt, Syria, Pakistan, Bangladesh). Both Western capitalism and Marxist socialism were rejected as being part of the problem rather than the solution, as having failed to redress widespread poverty and maldistribution of wealth. Capitalism was regarded as the system of special interests and new elites which produced a society more driven by materialism and conspicuous consumption than concern for equity and social justice. Marxism was dismissed as a godless alternative which struck at the heart of religion, substituting the material for the spiritual. Young people in particular found themselves in a world of shattered dreams which offered a dim future. In many countries idealism, study, and hard work were rewarded by unemployment or underemployment, housing shortages, and a lack of political participation which increased the sense of frustration and hopelessness.[6]

Socioculturally and psychologically, modernization was seen as a legacy of European colonialism perpetuated by Western-oriented elites who imposed and fostered the twin processes of Westernization and secularization. Just as dependence on Western models of development was seen as the cause of political and military failures, so, too, some

Muslims charged, blind imitation of the West and an uncritical Westernization of Muslim societies which some called the disease of "westoxification" led to a cultural dependence which threatened the loss of Muslim identity. The resultant process of secular, "valueless" social change was identified as the cause of sociomoral decline, a major contributor to the breakdown of the Muslim family, more permissive and promiscuous societies, and spiritual malaise.

The psychological impact of modernity, and with it rapid sociocultural change, cannot be forgotten. Urban areas had undergone physical and institutional changes, so that both the skylines and the infrastructure of cities were judged to be modern by virtue of their Western profile and façade. To be modern was to be Western in dress, language, ideas, education, behavior (from table manners to greetings), architecture, and furnishings. Urban areas became the primary locations and showplaces for work and living. Modern governments and companies as well as foreign advisers and investors focused on urban areas, so that the results of modernization only trickled down to rural areas. Rapid urbanization therefore meant the migration of many from outlying villages and towns. The hopes of the poor for a better life were often undermined by the realities of poverty in urban slums and shantytowns. Psychological as well as physical displacement occurred. Loss of village, town, and extended family ties and traditional values were accompanied by the shock of modern urban life and its Westernized culture and mores. Many, swept along in a sea of alienation and marginalization, found an anchor in religion. Islam offered a sense of identity, fraternity, and cultural values that offset the psychological dislocation and cultural threat of their new environment. Both the poor in their urban neighborhoods, which approximated traditional ghettos in the midst of modern cities, and those in the lower middle class who took advantage of the new educational and job opportunities of the city and thus experienced culture shock more profoundly and regularly, found a welcome sense of meaning and security in a religious revivalism. Islamic organizations' workers and message offered a more familiar alternative which was consistent with their experience, identified their problems, and offered a time-honored solution.

Finally, American ignorance of and hostility toward Islam and the Middle East, often critiqued as a "Christian Crusader" mentality influenced by Orientalism and Zionism, were blamed for misguided U.S. political–military policies: support for the "un-Islamic" Shah of Iran, massive military and economic funding of Israel, and the backing of an unrepresentative Christian-controlled government in Lebanon.

These crises and failures reinforced a prevailing Muslim sense of inferiority, the product of centuries of European colonial dominance, which left a legacy of both admiration of Western power, science, and technology and resentment of Western dominance, penetration, and exploitation.[7] The failures of the modern experience stood in sharp contrast to an Islamic ideal which linked the faithfulness of the Islamic community with worldly success, as witnessed by the memory of a past history in which Islam was a dominant world power and civilization.

From Failure to Success

During the seventies Islamic politics seemed to explode on the scene, as events in the Middle East (the Egyptian–Israeli war and the Arab oil embargo of 1973, as well as the Iranian Revolution of 1978–79) shocked many into recognition of a powerful new force that threatened Western interests. Heads of state and opposition movements appealed to Islam to enhance their legitimacy and popular support; Islamic organizations and institutions proliferated.

In 1973 Egypt's Anwar Sadat initiated a "holy war" against Israel. In contrast to the 1967 Arab–Israeli war which was fought by Gamal Abdel Nasser in the name of Arab nationalism/socialism, this war was fought under the banner of Islam. Sadat generously employed Islamic symbols and history to rally his forces. Despite their loss of the war, the relative success of Egyptian forces led many Muslims to regard it as a moral victory, since most had believed that a U.S.-backed Israel could not be beaten.

Military vindication in the Middle East was accompanied by economic muscle, the power of the Arab oil boycott. For the first time since the dawn of colonialism, the West had to contend with and acknowledge, however begrudgingly, its dependence on the Middle East. For many in the Muslim world the new wealth, success, and power of the oil-rich countries seemed to indicate a return of the power of Islam to a community whose centuries-long political and cultural ascendence had been shattered by European colonialism and, despite independence, by second-class status in a superpower-dominated world. A number of factors enhanced the Islamic character of oil power. Most of the oil wealth was located in the Arab heartland, where Muhammad had received the revelation of the Quran and established the first Islamic community-state. The largest deposits were found in Saudi Arabia, a self-styled Islamic state which had asserted its role as keeper of the holy cities of Mecca and Medina, protector of the annual pilgrimage (*hajj*), and leader and benefactor of the Islamic world. The House of Saud used its oil wealth to establish numerous international Islamic organizations, promote the preaching and spread of Islam, support Islamic causes, and subsidize Islamic activities undertaken by Muslim governments.

No event demonstrated more dramatically the power of a resurgent Islam than the Iranian Revolution of 1978–79. For many in the West and the Muslim world, the unthinkable became a reality. The powerful, modernizing, and Western-oriented regime of the Shah came crashing down. This was an oil-rich Iran whose wealth had been used to build the best-equipped military in the Middle East (next to Israel's) and to support an ambitious modernization program, the Shah's White Revolution. Assisted by Western-trained elites and advisers, the Shah had governed a state which the United States regarded as its most stable ally in the Muslim world. The fact that a revolution against him and against the West was effectively mounted in the name of Islam, organizing disparate groups and relying upon the mullah–mosque network for support, generated euphoria among many in the Muslim world and convinced Islamic activists that these were lessons for success to be emulated. Strength and victory would belong to those who pursued change in the name of Islam, whatever the odds and however formidable the regime.

For many in the broader Muslim world, the successes of the seventies resonated with an idealized perception of early Islam, the Islamic paradigm to be found in the time of the Prophet Muhammad, the Golden Age of Islam. Muhammad's successful union of disparate tribal forces under the banner of Islam, his creation of an Islamic state and society in which social justice prevailed, and the extraordinary early expansion of Islam were primal events to be remembered and, as the example of the Iranian Revolution seemingly verified, to be successfully emulated by those who adhered to Islam. Herein lies the initial attraction of the Iranian Revolution for many Muslims, Sunni and Shii alike. Iran provided the first example of a modern Islamic revolution, a revolt against impiety, oppression, and injustice. The call of the Ayatollah Khomeini for an Islamic revolution struck a chord among many who identified with his message of anti-imperialism, his condemnation of failed, unjust, and oppressive regimes, and his vision of a morally just society.

By contrast, the West stood incredulous before this challenge to the Shah's "enlightened" development of his seemingly backward nation, and the resurrection of an anachronistic, irrational medieval force that threatened to hurtle modern Iran back to the Middle Ages. Nothing symbolized this belief more than the black-robed, bearded mullahs and the dour countenance of their leader, the Ayatollah Khomeini, who dominated the media, reinforcing in Western minds the irrational nature of the entire movement.

The Ideological Worldview of Islamic Revivalism

At the heart of the revivalist worldview is the belief that the Muslim world is in a state of decline. Its cause is departure from the straight path of Islam; its cure, a return to Islam in personal and public life which will ensure the restoration of Islamic identity, values, and power. For Islamic political activists Islam is a total or comprehensive way of life as stipulated in the Quran, God's revelation, mirrored in the example of Muhammad and the nature of the first Muslim community-state, and embodied in the comprehensive nature of the Sharia, God's revealed law. Thus the revitalization of Muslim governments and societies requires the reimplementation of Islamic law, the blueprint for an Islamically guided and socially just state and society.

While Westernization and secularization of society are condemned, modernization as such is not. Science and technology are accepted, but the pace, direction, and extent of change are to be subordinated to Islamic belief and values in order to guard against the penetration of Western values and excessive dependence on them.

Radical movements go beyond these principles and often operate according to two basic assumptions. They assume that Islam and the West are locked in an ongoing battle, dating back to the early days of Islam, which is heavily influenced by the legacy of the Crusades and European colonialism. and which today is the product of a Judaeo-Christian conspiracy. This conspiracy is the result of superpower neocolonialism and the power of Zionism. The West (Britain, France, and especially the United States) is blamed for its support of un-Islamic or unjust regimes (Egypt, Iran, Lebanon) and also for its biased support for Israel in the face of Palestinian displacement. Violence against such governments and their representatives as well as Western multinationals is legitimate self-defense.

Second, these radical movements assume that Islam is not simply an ideological alternative for Muslim societies but a theological and political imperative. Since Islam is God's command, implementation must be immediate, not gradual, and the obligation to do so is incumbent on all true Muslims. Therefore individuals and governments who hesitate, remain apolitical, or resist are no longer to be regarded as Muslim. They are atheists or unbelievers, enemies of God against whom all true Muslims must wage jihad (holy war).

Exporting the Revolution

The success of the Iranian Revolution fired the imagination of the Muslim world and made Muslim governments tremble. Postrevolutionary

Iran influenced Islamic activists throughout the world. In the aftermath of the revolution, delegations of Muslim leaders came from North America, the Middle East, and Southeast Asia to Teheran to congratulate Khomeini. Sunni as well as Shii activist organizations, extending from Egypt (the moderate Muslim Brotherhood and radical al-Jihad) to Malaysia (ABIM, or the Malaysian Youth Movement, and the militant PAS) drew inspiration from the example of Iran. The Iranian Revolution provided lessons "to awaken Muslims and to restore their confidence in their religion and their adherence to it, so that they may assume the reins of world leadership of mankind once again and place the world under the protection of the esteemed Islamic civilization."[8] The Iranian Revolution served as a reminder that Islam is a comprehensive way of life which regulates worship and society: "It is religion and state, governance and politics, economics and social organization, education and morals, worship and holy war."[9]

Long-quiescent Shii minority communities in Sunni-dominated states like Saudi Arabia, the Gulf, and Pakistan aggressively asserted their Shii identity and rights and were emboldened to express discontent with ruling regimes. At the same time, threatened Muslim rulers increasingly branded their Islamic opposition as Khomeini-like or accused Iran of exerting undue influence on their domestic politics.

Gulf rulers were particularly nervous about the appeal of Iran's revolutionary example and rhetoric. In Iraq, where Shii constitute 60 percent of the population, the government of Saddam Hussein, a nominally Sunni Muslim ruler, was shaken by eruptions in the Shii cities of Karbala, Najaf, and Kufa (June 1979). Khomeini denounced Saddam Hussein as an atheist and called for the overthrow of his regime. Saddam Hussein countered by vilifying Khomeini and appealing to Iran's minority Arab population to revolt. The government suspected Iranian influence within Iraq's Shii activist groups—in particular, in the Islamic Call Society (al-Dawa) and the newly formed (1979) Mujahidin.[10] The Ayatollah Muhammad Baqir al-Sadr, one of Iraq's most prominent and influential Shii activist clerics, who had befriended Khomeini during his exile in Iraq, welcomed Iran's revolution and Khomeini's Islamic government. Saddam Hussein acted decisively. Shii leaders were arrested. Baqir al-Sadr, who had declared Iraq's Baathist regime un-Islamic and forbidden any dealings with it, was executed (April 1980), and al-Dawa was outlawed. In an atmosphere in which both Iraq and Iran played upon centuries-long Arab–Persian and Sunni–Shii rivalries and hostilities, the situation deteriorated. In September 1980 Iraq invaded Iran, initiating a war that would last eight years.

Khomeini was particularly critical of the Saudi and Gulf governments.

He denounced them as "un-Islamic" monarchies, disdainfully character-izing their military and economic ties with the United States as "Ameri-can Islam." Audiotapes of Khomeini and revolutionary leaflets were smuggled into these Sunni-dominated states, and daily Arabic broad-casts from Teheran were explicit in their critique and their agenda:

> The ruling regime in Saudi Arabia wears Muslim clothing, but it actually represents a luxurious, frivolous, shameless way of life, robbing funds from the people and squandering them, and engaging in gambling, drink-ing parties, and orgies. Would it be surprising if people follow the path of revolution, resort to violence and continue their struggle to regain their rights and resources?[11]

The worst fears of Gulf rulers and of the West—that Iran's Revolution would prove contagious—seemed to be coming true in the first years after the revolution. In November 1979 Saudi Arabia was rocked by two explosive events. On November 20, as Muslims prepared to usher in the fifteenth century of Islam, the Grand Mosque at Mecca was seized and occupied for two weeks by Sunni militants who denounced the Saudi monarchy. Khomeini's accusation that Americans had been behind the mosque seizure led to attacks against American embassies and the de-struction by fire of the American embassy in Islamabad, Pakistan.

While still reeling from the seizure of the Grand Mosque, on Novem-ber 27 the house of Saud saw riots break out among the 250,000 Shii Muslims in the oil-rich Eastern Province (Al-Hasa) where Shii constitute 35 percent of the population.[12] Pent-up emotions and grievances among Shii, who felt discriminated against by their Sunni rulers and cheated by an unfair distribution of oil wealth and government services, had ex-ploded earlier in the year in response to Iran's revolution and the trium-phant return of Khomeini.

Events in the early eighties did nothing to lessen concern among governments in the Gulf, the wider Muslim world, and the West. State-ments by the ruling ayatollahs in Iran, who called for an aggressive, expansionist policy, exacerbated the situation. President Khamenei called upon prayer leaders from forty countries to turn their mosques into houses of "prayer, cultural and military bases [to] . . . prepare the ground for the creation of Islamic governments in all countries."[13] The Iran–Iraq war, which officially began with Iraq's invasion of Iran on September 22, 1980, inflamed relations between Iran and its neighbors. The Gulf states organized the Gulf Cooperation Council (GCC) and threw their substantial financial support to Iraq. Khomeini called upon the GCC to "return to the lap of Islam, abandon the Saddam Hussein regime in Baghdad, and stop squandering the wealth of their peoples."[14]

During the same period Bahrain and Kuwait were also threatened by Shii unrest.[15] In 1981 the government of Bahrain foiled an Iranian-inspired coup by the Shii Front for the Liberation of Bahrain. Kuwait, 30 percent of whose population is Shii, was hit by car bombings of the American and French embassies (1983) and was forced to crack down on Shii unrest in 1987 and 1989.

Iran frequently used the annual *hajj* to Mecca to propagate its revolutionary message. The Ayatollah Khomeini and other senior clerics rejected the Saudi claim to be the keepers of the holy sites and maintained that the *hajj* had a rightful political dimension. Iranian pilgrims, displaying posters of Khomeini and chanting slogans against the United States, the Soviet Union, and Israel, clashed with Saudi security in June 1982. The tensions continued during subsequent years and climaxed in 1987 when more than four hundred people were killed in a confrontation between Iranian pilgrims and Saudi security forces.

Despite these sporadic disturbances and government fears of massive unrest, Iran's export of its revolution proved in the end to be surprisingly unsuccessful in rallying Iraqi Shii as well as the populations of the Gulf states. By and large, most Iraqi (Arab) Shii were swayed more by nationalist than religious ties to their (Persian) coreligionists in Iran. Pockets of Shii militancy in the Gulf states did not translate into significant revolutionary movements. To thwart their growth, governments successfully used a combination of carrot and stick, addressing socioeconomic grievances while increasing security and imprisoning and deporting dissidents.

Lebanon provides the clearest and boldest example of the direct impact of Iran's Revolution. After the revolution Iran sent a contingent of its Revolutionary Guards to Lebanon who influenced the development there of Shii militant organizations, in particular Hizbullah and al-Jihad, supplying training, material, and money. In contrast to the more moderate Shii AMAL, which fought for a more representative reapportionment of power within the existing state structure, the goal of Iranian-inspired organizations was the eventual creation of an Islamic state. Iran was suspected of attempting to further that end through everything from bombings of Western embassies to car bomb attacks and the taking of hostages.

From the Periphery to the Center: Mainstream Revivalism

While the exploitation of Islam by governments and by extremist organizations has reinforced the secular orientations of many Muslims and cynicism in the West, a less-well-known and yet potentially far-reaching social transformation has also occurred in the Muslim world. In the nineties Islamic revivalism has ceased to be restricted to small, marginal organizations on the periphery of society and instead has become part of mainstream Muslim society, producing a new class of modern-educated but Islamically oriented elites who work alongside, and at times in coalitions with, their secular counterparts. Revivalism continues to grow as a broad-based socioreligious movement, functioning today in virtually every Muslim country and transnationally. It is a vibrant, multifaceted movement that will embody the major impact of Islamic revivalism for the foreseeable future. Its goal is the transformation of society through the Islamic formation of individuals at the grass-roots level. *Dawa* (call) societies work in social services (hospitals, clinics, legal-aid societies), in economic projects (Islamic banks, investment houses, insurance companies), in education (schools, child-care centers, youth camps), and in religious publishing and broadcasting. Their common programs are aimed at young and old alike.

A more pronounced Islamic orientation is now to be found among the middle and lower classes, educated and uneducated, professionals and workers, young and old, men, women, and children. A new generation of Islamically oriented leaders have appeared in Egypt, the Sudan, Tunisia, Jordan, Iran, Malaysia, Kuwait, Saudi Arabia, and Pakistan. Islamic activists have become an accustomed part of the political process, participating in national and local elections, scoring an impressive victory in Algeria's municipal elections, emerging as the chief opposition parties or groups in Egypt, Tunisia, and Jordan, and serving in cabinet positions in the Sudan, Jordan, Pakistan, Iran, and Malaysia.

For many who have nurtured visions of Islamic activism as the work of a small band of radicals and terrorists, the reality of Islam's strength, its expanding spheres of influence in mainstream politics and society, are both bewildering and menacing. How is one to make sense of the growth and prominence of Islamic activism in much of the Muslim world? What are the roots of the Islamic resurgence? Is Islamic fundamentalism simply a form of religiously motivated terrorism and extremism? If so, why does it enjoy such widespread support? Why is Islam so prominent in Muslim politics? Where and how has Islam been reasserted in politics and society? Why has Islamic revivalism proved to be so strong in those countries long regarded as among the most Westernized—Egypt, Lebanon, Iran, and Tunisia? How could Saddam Hussein, the most secular and un-Islamic of leaders, have appealed to Islam in the 1991 Gulf War? Finally, to what extent is there an "Islamic threat"?

Saddam Hussein's attempt to rally populist Muslim opinion for yet another holy war between Islam and the West is but the most recent episode in a long history of confrontation between Islam and the West. Muslim–Christian relations have often seemed a history of confrontation, with each side regarding the other as a historic threat from the rise of Islam to the present. Early encounters and confrontations, theological and political, provide the images and folklore which sustain the mutual stereotypes, images, and suspicions that continue to fuel fears and biases and perpetuate a vision of Islam against the West or of the West against Islam.

2

Islam and the West: Roots of Conflict, Cooperation, and Confrontation

Despite common theological roots and centuries-long interaction, Islam's relationship to the West has often been marked by mutual ignorance, stereotyping, contempt, and conflict. Ancient rivalries and modern conflicts have so accentuated differences as to completely obscure the shared theological roots and vision of the Judaeo-Christian–Islamic tradition. Both sides have focused solely on and reinforced differences, and have polarized rather than united these three great interrelated monotheistic traditions.

Islam's early expansion and success constituted a challenge theologically, politically, and culturally which proved a stumbling block to understanding, and a threat to the Christian West. Both Islam and Christianity possessed a sense of universal message and mission which in retrospect were destined to lead to confrontation rather than mutual cooperation. Because of a long history in which Christendom often vilified the Prophet and grossly distorted Islam, and a recent history in which Islam has often been equated with radicalism and terrorism, some understanding of Islam is necessary before moving on to the history of relations between the West and the Muslim world. Moreover, some awareness of the Quran, the Prophet Muhammad, and the early Islamic period is indispensable for understanding Islam, since they have continued to provide a paradigm for emulation by Muslims and Islamic movements in every age. As we struggle to move beyond stereotypes and the tendency to confine our images of Islam and Muslims to demonstrators shouting "Death to America" or "Holy war against the infidels," it is important to remember that men and women of every race and color, social class, and educational background across the world and down through the ages have found in Islam a faith that nourishes and transforms their lives, one that offers a sense of community, solidarity, and peace.

The Origins and Nature of Islam

Although the origins of the Muslim community date back to Muhammad and the seventh century of the common era, Muslims, like Jews and Christians, also trace the origins of their religious tradition (Islam) to the one true God (Allah, "the God") through a long line of prophetic messengers. Thus Muslims emphasize that the Judaeo-Christian tradition is more accurately the Judaeo-Christian–Islamic tradition because all three are children of Abraham, the first prophet to receive God's revelation. They share an Abrahamic faith with its common belief in God, prophets, revelation, a divinely mandated community, and moral responsibility.[1] While Jews and Christians trace their lineage through Isaac to Abraham and Sarah, Muslims do so through Ishmael, the first-born son of Abraham and his servant Hagar.

Much as the sacred story of the Hebrew people records the triumph of a monothesim revealed to Adam, Abraham, and Moses in a polytheistic world, so Islamic history recounts a similar process of uncompromising monotheism, prophecy, and divine revelation within a tribal, polytheistic social setting.[2] As Jews have the Hebrew scriptures or Torah, and Christians the Bible, so Muslims possess their sacred scripture or book, the Quran. Muslims believe that God first sent His revelation to the Jews and then to the Christians, but that the revelation became distorted through human intervention and interpolation of the scriptures, resulting in such beliefs as the incarnation, crucifixion, death, and resurrection of Jesus, and the doctrine of redemption. The Hebrew Bible or Old Testament and the Christian New Testament are believed to be flawed versions of the original, pristine divine revelation. God subsequently sent down his revelation one more time through Muhammad, the last and final prophet. This, then, is the basis for the Muslim belief that the Quran, which Muslims view as the perfect, complete, and literal word of God, supersedes Jewish and Christian scriptures.

Muhammad: Messenger of God

The Quran and the Prophet Muhammad provide the sacred sources and guidance for the development of faith in both past and present. As the followers of Muhammad turned to him during his lifetime, so today devout Muslims across the world look to revelation and the teachings of the Prophet for direction in their lives.

Born in Arabia (roughly, modern-day Saudi Arabia) in 570, Muhammad ibn Abdullah (570–632) had a profound religious experience at the age of forty which transformed his life and initiated a community that some fourteen centuries later would become the second largest of the world's religions, spanning the globe and numbering approximately one billion adherents. In comparison to most prophets or founders of the world's great religious traditions, whose lives remain undocumented, a significant number of early documents provide information on the Prophet's life and work: the Quran, prophetic traditions (Hadith, reports about what the Prophet said and did), and an early biography by ibn-Ishaq (d. circa 768).[3] However, we know little of Muhammad's early years. He was orphaned as a child and raised by relatives. Muslim tradition tells us that as a young man of twenty-five he married a wealthy widow. Khadija was the owner of the caravan for which he was business manager and was some fifteen years older than Muhammad. Religiously inclined, Muhammad often retreated to a quiet place to reflect and meditate. In 610 C.E., on a night which Muslims commemorate as the "Night of Power and Excellence," Muhammad the caravan leader became Muhammad the Messenger of God, receiving the first of many revelations through the angel Gabriel: "Recite in the name of your Lord who has created, created man out of a germ-cell. Recite for your Lord is the Most Generous One who has taught by the pen, taught man what he did not know!" (Quran 96:1–5). Coming over the period from 610 to 632, the revelations would be fully compiled after the Prophet's death and constitute Islam's sacred scripture, the Quran.

Muslim tradition portrays an initially confused and somewhat reluctant prophet who, like the biblical Hebrew prophets, was overwhelmed by the experience—confused as to its meaning and worried about the reception of his special claims by others. As the history of the prophetic tradition demonstrates, those called to be a warner or messenger from God do not have an easy life. Prophets who claim to be a divinely inspired conscience of society, who denounce the waywardness and infidelity of the community, and who challenge the establishment and prevailing culture often faced derision, rejection, and persecution. Muhammad was no exception.

For ten years he preached a message of religious and social reform in Mecca. Muhammad and the Quran proclaimed one true God, rejecting the prevailing polytheism of Arabia, and denounced social injustice. Muhammad did not claim to bring a new religion but to purify and

restore the one true religion of Abraham. His message was one of reformation and restoration of a wayward and faithless community. Like
Amos and Jeremiah before him, Muhammad was a messenger sent from
God who denounced the impiety of his society and called upon his
hearers to repent and obey God, for the final judgment was near: "Say:
'O men, I am only for you a warner.' Those who believe and do deeds of
righteousness—theirs shall be forgiveness and generous provision. And
those who strive against our signs to avoid them—they shall be inhabitants of Hell" (Quran 22:49–50).

Muhammad called upon Meccan society to worship the one true God
and to abandon its polytheistic cult and practices. Arabia was no
stranger to monotheism. However, while there were Jewish and Christian communities as well as indigenous Arabian monotheists (Hanifs), a
vast array of tribal gods and goddesses dominated Arabian society. Muhammad preached a return to the faith of Abraham: belief in the one
true God, the creator, sustainer, and judge of the universe. Muhammad
and the Quran taught that human beings were accountable and that they
would be judged and eternally rewarded or punished for their actions on
Judgment Day. The call to Islam was a call to turn away from the path of
unbelief and return to the straight path (Sharia) or law of God. This
conversion meant membership in a community dedicated to the worship
of the one true God, the implementation of His will, and thus the
creation of a socially just community.

The Quranic message was a challenge to the prevailing sociopolitical
as well as religious order. Mecca was a center not only of religious
pilgrimage but also of trade and commerce, caught in the transition from
a semi-Bedouin tribal to a commercial urban society. Obedience to God
and his Prophet, a brotherhood of believers, alms for the poor, and
struggle (jihad) against oppression were prescribed. The Quran condemned the exploitation of the poor, orphans, and women; forbade
corruption, fraud, cheating, false contracts in business, the flaunting of
wealth and arrogance; and prescribed strict punishments for slander,
stealing, murder, the use of intoxicants, gambling, and adultery. Muhammad's claim to prophetic authority, his denunciation of injustices in
Meccan society, and insistence that all believers belonged to a single
universal community undermined tribal political authority. His rejection
of polytheism seriously threatened the economic interests of the Mec-
cans who controlled the Kaaba, a sacred shrine that housed the tribal
idols and was the site of a great annual pilgrimage and fair, the source of
Meccan religious prestige and revenue.

After ten years Muhammad enjoyed limited success. By worldly standards he would have been regarded as a failure. Although protected by

his influential uncle, abu-Talib, and by his kinsman of the Banu Hashim, he himself lacked the power and prestige to overcome widespread opposition from the Meccan aristocracy, led by the Quraysh, the dominant merchant/trader clan in Mecca. By 619, with the passing of both his wife and his uncle, Muhammad had lost the pillars of his personal support and protection and was even more alone and vulnerable. His small band of followers was increasingly persecuted by the Meccan aristocracy, who regarded Muhammad's claim to prophecy and reformist agenda, with its implicit criticism of the political and socioeconomic status quo, as a challenge to their leadership and interests. For these reasons, when invited by leaders of the neighboring city of Medina, an agricultural oasis, to serve as its arbitrator, he and a band of his Meccan followers migrated in 622 and established there the first Muslim community (*ummah*).

The Islamic Community

The importance of the *hijra* (hegira), the emigration of Muhammad and his companions from Mecca to Medina, cannot be overestimated. The *hijra* constitutes the beginning of the Muslim community and was of such religious and historical significance that it became the basis for reckoning Muslim history, the first year of the Islamic calendar. Muhammad's move to Medina signaled a major transformation in history from a pre-Islamic pagan past to a divinely guided and centered world in which tribal kinship was to be superseded by membership in a community (*ummah*) bound together by common religious belief.

The community at Medina consisted of many clans, including several Jewish clans. Over the course of time Muhammad was able to consolidate his power and authority, converting the pagan clans to Islam. Muhammad was at last both Prophet and leader of a community.[4] Islam did not know a sharp cleavage between sacred and profane, between religion and society, individual and community. Islam's orientation was symbolized in the recurrent Quranic command, "Obey God and His Prophet." The Prophet received and transmitted a revelation that both guided and responded to the dynamics of the community's history. At the same time he oversaw and governed its affairs, serving as its political and military leader, judge and social reformer. Religion was integral to the leadership, life, and fabric of society, providing norms for worship (duties to God) and social life (duties to society). If Islam means submission to the will of God, the Muslim is one who submits, that is, follows or actualizes God's will in both individual and community life.

Medina reflected the integral relationship of religion and state in Islam, an ideal that influenced the development of the Islamic community both then and now. The state was led by Muhammad, the Prophet of God, and guided by divine revelation. Muhammad exercised executive, legislative, and judicial roles as head of state. He oversaw domestic and foreign affairs, the military, and the collection of taxes, and he settled disputes. The Islamic community was obliged not only to follow but also to spread God's word and rule. Muhammad combined diplomacy and military action. He returned in triumph to Mecca and by the time of his death in 632 had consolidated the disparate tribes of Arabia.

Muhammad brought more than simply a new synthesis or interpretation of existing religious ideas and customs (Arabian, Jewish, Christian). He fashioned a new order and community, a religious and political (more accurately, a religiopolitical) community rooted in and united by a religious vision or bond.[5] Old ideas and institutions were adopted and transformed in light of Islamic norms, as a new sense of identity, solidarity, community, and authority were fashioned. At the heart of the movement was a "new religion," a special understanding of the meaning and implications of Islam's monotheistic vision and way of life: the oneness of a transcendent and all-powerful God; His all-encompassing will for creation, which impacted on personal and public life; the requirement of obedience to that will; the mission of each Muslim as an individual and as a member of a religious community to realize and spread God's rule; and the completeness and correctness of the Quranic revelation and the finality of Muhammad's prophethood. It was this vision that was to transform the tribes of Arabia and bring about a major world historical and cultural transformation.

Islamic Government and Society

Muhammad's death precipitated a crisis of leadership in the community whose resolution would see the creation of the caliphate and also sow the seeds for the later emergence of Islam's two major branches, Sunni and Shii Islam. With Muhammad's death, the Muslim community faced the selection of a successor. Two opinions emerged. The majority believed that Muhammad had not designated a successor and accepted the selection of his successor (caliph) by Muhammad's senior companions. The caliph was to be the political leader of the community with no claim to the prophetic mantle of the Prophet. However, a minority believed that Muhammad had designated Ali, who, as his cousin and son-in-law, was the senior male in Muhammad's family. For these followers of Ali (Shii or "partisans" of Ali), leadership of the Islamic community was to stay within the house of the Prophet. Ali and his descendents were to be the religiopolitical leaders (Imams) of the community. Though not a prophet, the Shii Imam, in contrast to the office of caliph, enjoyed a very special religious status as a religiously inspired and sinless leader.

The majority Sunni opinion prevailed. The Islamic community was organized and governed under a caliphal state which was quickly transformed into a central, imperial, dynastic caliphate.[6] The first six centuries of Islam can be neatly divided into three major periods: the Four Rightly Guided Caliphs at Medina (632–661), so called because, in conjunction with the era of Muhammad, their rule is regarded as the ideal formative and normative period of Islam; the Umayyad Caliphate (661–750) at Damascus; and the Abbasid Caliphate (750–1258) at Baghdad.

The worldview that prevailed during the caliphate period was deeply indebted to religion. While rulers relied upon their armies for security and conquest, Islam provided the ideological framework for state and society, a source of legitimacy and authority. Whatever the character of the ruler or how he came to power, all sought legitimacy in their caliphal title as successors to the Prophet, as the "commander of the believers," whose task it was to defend and spread the faith and assure that society was governed by God's law. The world was divided into Islamic territory (the *dar al-Islam,* abode of peace) and the non-Islamic world (the *dar al-harb,* abode of war). Citizenship, taxation, and issues of war and peace were determined by religious belief. Muslims enjoyed full citizenship and paid certain taxes, while Jews and Christians (People of the Book) were designated "protected people" and paid a special poll or head tax, in exchange for which they were to be defended by the Muslim military.

The Spread and Conquests of Islam

Within one hundred years of the death of the Prophet Muhammad, the successors (caliphs) of Muhammad had established an empire greater than Rome at its zenith. The shock to the international order and more specifically to Christendom was incalculable. That the tribes of Arabia could be united, let alone spill out of Arabia, overcome the Byzantine (Eastern Roman) and Persian (Sassanid) empires, and by the end of a century create an Islamic caliphate extending from North Africa to India, seemed unthinkable.

There are many reasons for the rapidity and success of Arab expansion: the mutual exhaustion of the Byzantine and Persian empires after years of war, the dissatisfaction of indigenous populations with their imperial rulers, the skill of Bedouin warriors, the lure of booty. However, the critical factors were the rise of the state and the role of Islam in uniting disparate tribes and providing a greater sense of purpose and meaning:

Islam . . . provided the ideological underpinnings for this remarkable breakthrough in social organization. . . . In this sense, the conquests were truly an *Islamic* movement. For it was Islam—the set of religious beliefs preached by Muhammed, with its social and political ramifications—that ultimately sparked the whole integration process and hence was the ultimate cause of the conquests' success.[7]

The perception or belief that this was indeed an Islamic movement remains prevalent today among many Muslims as a source of inspiration and emulation. Because of the tendency in our secular age to downplay religion as a major factor in sociopolitical development, it is useful to bear in mind a similar comment regarding the central importance of Islam in the conquests from a work by two Western, non-Muslim historians.

Even the secular historian . . . must regard Islam as the decisive factor in the expansion of the Arabs. That bedouin tribes, which constantly for centuries had engaged in warfare with each other and were known to prize their independence, should have suddenly placed themselves obediently under the order of Muslim commanders is inconceivable apart from Islam. It was Islam that provided the necessary rallying cry and instilled in the bedouin warriors a sense that they were fighting in a grand cause. Whatever may have been the original material motives of the bedouin . . . [they found themselves] caught up in a movement greater than anything they had dreamt of, a movement not of their own making, which they could explain only in terms of a divine intervention in human affairs.[8]

During subsequent centuries Islam would spread across much of the world. While the centralized caliphate disintegrated, it was replaced by individual states (sultanates) which extended from Africa to Southeast Asia, from Timbuktu to the southern Philippines.[9] In addition, great Muslim cities existed in what today are the Central Asian republics of the former Soviet Union, China, Eastern Europe, Spain, southern Italy, and Sicily.

Jihad

The rapid spread of Islam proved a double threat to Christendom, both religious and political. The armies and merchants of Islam were its missionaries, bringing both Muslim faith and imperial rule. Islam was used to bind together, inspire, and mobilize the tribes as well as to provide a rationale for expansion and conquest. The Quranic notion of jihad, striving or self-exertion in the path of God, was of central significance to Muslim self-understanding and mobilization. The term *jihad* has a number of meanings which include the effort to lead a good life, to make society more moral and just, and to spread Islam through preaching, teaching, or armed struggle. Muslim jurists distinguished ways "in which the duty might be fulfilled: by the heart, by the tongue, by the hands and by the sword."[10]

In its most generic meaning, "jihad" signifies the battle against evil and the devil, the self-discipline (common to the three Abrahamic faiths) in which believers seek to follow God's will, to be better Muslims. It is the lifelong struggle to be virtuous, to be true to the straight path of God. This is the primary way in which the observant Muslim gives witness to or actualizes the truth of the first pillar of Islam in everyday life. The spread of Islam through "tongue" and "hands" refers to the Quranically prescribed obligation of the Muslim community "to enjoin good and forbid evil" (3:110). Finally, "jihad" means the struggle to spread and to defend Islam. Just as the example of the Prophet offers a paradigm and the basis for the fusion of religion and state, so too Muhammad's movement readily supplies the model for all Islamic movements in their struggle to reform society and the world. The world is a battleground on which believers and unbelievers, the friends of God and the enemies of God or followers of Satan, wage war: "The believers fight in the way of God, and the unbelievers fight in the idol's way. Fight you therefore against the friends of Satan" (4:76). The mission of the Islamic community is to spread the rule or abode of Islam globally much as Muhammad and his followers expanded Islamic rule through preaching, diplomacy, and warfare, and to "defend" it. Islamic law stipulates that it is a Muslim's duty to wage war against polytheists, apostates, and People of the Book who refuse Muslim rule, and those who attack Muslim territory. To die in battle is the highest form of witness to God and to one's faith. The very Arabic word for martyr (*shahid*) comes from the same root as the profession of faith (*shahada*). As in Christianity, the reward for martyrdom is paradise.

Islamic Civilization

As Muslims spread their rule and faith, they proved to be great learners as well as doers. Politically, Muslim rulers recognized their own limitations and the advanced development of many of the kingdoms and cultures that their armies conquered. Local institutions, ideas, and personnel were assimilated or retained, and adopted or adapted to Islamic norms, as Muslim masters learned from their more advanced subjects. Great libraries and translation centers were established; the great books of science, medicine, and philosophy of the West and the East were collected and translated, often by Christian and Jewish subjects, from

Greek, Latin, Persian, Coptic, Syriac, and Sanskrit into Arabic. Thus the best works of literature, science, and medicine were made more accessible.

The age of translation was followed by a period of great creativity, as a new generation of educated Muslim thinkers and scientists now built upon their received knowledge and made their own contributions to learning. "The process of Islamicizing the traditions had done more than integrate and reform them. It had released tremendous creative energies. The High Caliphal Period was one of great cultural florescence."[11] This was the age of great masters of philosophy and science: ibn-Sina (Avicenna), ibn-Rushd (Averroës), al-Farabi. Major urban centers of learning with vast libraries emerged in Cordova, Palermo, Nishapur, Cairo, Baghdad, Damascus, and Bukhara to eclipse a Europe mired in the Dark Ages. The political and cultural life of Muslims and non-Muslims alike in Islamic empires and states, despite tribal and religious differences, were brought within the framework of Islamic faith and Arabic language. New ideas and practices were Arabized and Islamized. Islamic civilization was the product of a dynamic, creative process of change in which Muslims borrowed freely from other cultures. It demonstrated a sense of openness and self-confidence that came from being masters not servants, colonizers rather than the colonized. In contrast to the twentieth century, Muslims then enjoyed a sense of control and security. They felt free to borrow from the West, since their identity and autonomy were not threatened by the specter of political and cultural domination. As they borrowed, so too they provided a legacy to the West. The earlier cultural traffic pattern was reversed when Europe, emerging from the Dark Ages, turned to Muslim centers of learning to reappropriate much of its lost heritage and to learn from Muslim advances in mathematics, medicine, and science.

Islamic Law: The Path of God

If the Muslim vocation was to follow or obey God's will, then knowing God's will was imperative. Whereas dogma or doctrine epitomized the essential statement of Christian belief, Islam, like Judaism, found its central expression in law.[12] Law rather than theology was the dominant discipline for defining or delineating faith. For Muslim jurists, God's revelation and prophetic example were the starting points for discerning and applying God's will to every aspect of life. Both the message of the Quran and the traditions of the Prophet reveal the comprehensiveness of Islam's way of life, its individual and corporate dimensions. Within several centuries of the death of the Prophet, Muslims had codified their way of life. Pious Muslims concerned about the unfettered powers of Muslim rulers and the infiltration and uncritical assimilation of foreign practices, sought to delineate God's law in order to preserve the true path of God and to limit the powers of the caliph.

Based on the Quran and the example of the Prophet and utilizing custom and reason, the work of individual jurists gave birth to schools (communities of scholars) of law which sprang up in many of the great cities of Islam: Medina, Mecca, Damascus, Baghdad, Kufa. Though united in purpose and based upon the same revealed sources, their conclusions often bore the mark of differing geographic contexts and customs as well as intellectual orientations. Of the many law schools which sprang up, several (the Hanafi, Maliki, Shafii, Hanbali, and Jafari) would survive and endure.

Islamic law provided the blueprint of the good society, the Islamic ideal. The Sharia or path of God was therefore a set of divinely revealed general principles, directives, and values from which human beings developed detailed rules and regulations which were in turn to be applied by judges (*qadis*) in Sharia courts.

The scope of Islamic law was comprehensive, including regulations which governed ritual and worship and defined the social norms of the community. Central to the faith are the five pillars, or basic duties, which are incumbent upon all believers.

    1. The Profession of Faith (*shahada*) which marks entrance into or membership in the Islamic community: "There is no God but the God and Muhammad is the messenger or prophet of God."

    2. Prayer (*salat*) five times each day at fixed times, and attendance at the Friday congregational prayer.

    3. Almsgiving (*zakat*), a two-and-a-half percent tithe on a Muslim's accumulated wealth, which is to be distributed to the poor not as charity but as a religious obligation of all Muslims toward the less fortunate brothers and sisters of the community.

    4. Fasting (*sawm*) from dawn to dusk during the month of Ramadan.

    5. Participating in the annual pilgrimage (*hajj*) to Mecca at least once in one's life, a duty incumbent upon all who have the health and economic resources to do so.

The five pillars combine a sense of individual responsibility, social awareness, and collective consciousness or membership in the broader community of Islam.

The social dimension of the law is embodied in a set of regulations or norms that govern family, criminal, contract, and international law. Here in particular one sees the impact of Islam on both personal and community life. A vast body of laws regulated marriage, polygamy or

polygyny, divorce, inheritance, theft, adultery, drinking, and issues of
war and peace.

While it possessed an underlying essential unity, Islamic law reflected
the diversity of the geographical contexts, with their differing customs,
in which it developed, as well as differences due to human interpretation
or judgment. Islamic law, then, was neither rigid nor closed but instead
manifested dynamism, flexibility, and diversity. It remained responsive
to new circumstances in the hands of legal experts (muftis) who served
as advisers to the courts. Their interpretations (legal opinions or decrees, *fatwas*) either on the finer points of law or on new situations or
questions, often guided the findings of the court. By the tenth century,
however, Islamic law did tend to become more fixed as many jurists
concluded that the essentials of God's law had been adequately delineated in legal texts. Thus there was a tendency to restrict substantive
interpretation (*ijtihad*) and instead emphasize the obligation to simply
follow or imitate (*taqlid*) Islamic legal texts. New practices or doctrines
were condemned as deviations (*bida*) from God's revealed law, unwar-
ranted innovation often regarded as akin to heresy. As a result, over
time the distinction between God's immutable law as found in revelation
and many of the legal regulations which were the product of fallible
human reasoning or local custom became blurred and forgotten. The
question of Islamic law and change would become a major issue in the
nineteenth and twentieth centuries, as Muslims responded to the impact
of modernization and development.

Islamic Mysticism and Spirituality

Given the prescriptive role of Islamic law in defining official Islam and
the tendency in recent years to focus on political Islam, it has been easy
to lose sight of the rich mystical and spiritual tradition (Sufism) which
has nourished the lives of Muslims and accounts for the effective spread
of Islam throughout much of the world. The formal, legal tradition, with
its specific guidelines and punishments, has always been accompanied by
an inner-directed quest and path. The letter of the law was offset by an
emphasis on the inner spirit or heart of Islamic faith and belief.

Sufism began as a reform movement.[13] For some pious Muslims the
splendor and wealth of the conquests, the transformation of the relative
simplicity of life in Arabia into the imperial court life of Damascus,
threatened the pristine faith and moral fiber of the community. The
kingdom of man, with its focus on this world, seemed to have overwhelmed and obscured the kingdom of God, the true focus and center of
Muslim life. In Sufi eyes, Islam triumphant had become an Islam endan-

gered. Calling for a refocusing on the truths of the next life rather than the pleasures and rewards of this life, the Sufis preached a message of simplicity and detachment from the things of this world. To this early austerity was added a strain of spirituality that emphasized love of God. Sufism offered an affective, devotional path to God whose selfless love and service to God complemented and sometimes challenged the more legalistic, scripturalist approach. Thus the good Muslim was not only one who followed God's will through observance of the law, but also the believer who sought through various means including recollection, and sometimes song and dance, to draw closer to the divine, to experience the presence of God.

The fusion of asceticism and devotionalism transformed Sufism from a relatively small urban elite movement into a broad-based populist movement whose communities attracted followers from all social classes and educational backgrounds. Groups of Sufis, gathered around a spiritual leader or master (a *pir* or shaykh), formed brotherhoods or orders. From the twelfth to the fourteenth centuries, Sufi brotherhoods were transformed from small voluntary associations into organized brotherhoods with an international network of centers spreading across the Muslim world. They became the great missionaries of Islam, responsible for its effective spread. In Africa and Southeast Asia Islam was spread primarily by Sufi brotherhoods and merchants rather than the armies of Islam. Sufism brought a message of Islam whose mystical doctrines and practices proved attractive to many and was open to linkages with local religious traditions and customs. Whereas official Islam often emphasized strict observance of the letter of the law, Sufism presented an alternative tradition flexible and open to assimilation and synthesis. Outside influences were absorbed from Christianity, Neoplatonism, Hinduism, and Buddhism. As Sufism became a mass movement, it came into conflict with the official Islam of the *ulama,* whose authority in the community was challenged by Sufi popularity and success as the *ulama* and Sufi *pirs* often vied for influence.

Islam and the West

The rapid rise and expansion of the Islamic empire and the flourishing of Islamic civilization posed a direct danger to Christendom's place in the world both theologically and politically. As Maxime Rodinson has observed, "The Muslims were a threat to Western Christendom long before they became a problem."[14]

The very theological similarities of Christendom and Islam put the two on a collision course. Each community believed that its covenant with God was the fulfillment of God's earlier revelation to a previous community that had gone astray. Each believed in the history of God's revelation and that its revelation and messenger marked the end of revelation and prophecy. Thus while Christians assumed a position of superiority and therefore had little problem with their supercessionist views toward Judaism, a similar attitude and claim by Muslims regarding Christianity was unthinkable and, more than that, a threat to the uniqueness and divinely mandated role of Christianity to be the sole representative of God and the only means to salvation. Islam was at best a heresy preached by a deluded or misguided prophet, and at worst a direct challenge to Christian claims and mission: "the combination of fear and ignorance produced a body of legends, some absurd and all unfair: Muslims were idolaters worshipping a false trinity, Muhammad was a magician, he was even a Cardinal of the Roman Church who, thwarted in his ambition to become Pope, revolted, fled to Arabia and there founded a church of his own."[15]

Both Christianity and Islam claimed a universal mission; each was a transnational community based upon common belief and a vocation to be an example to the nations of the world, the vehicle for the spread and triumph of God's kingdom. However, the challenge of Islam was not at the level of theological discourse and debate. The success of Muslim armies and missionaries was experienced as a force which seemed to come out of nowhere to challenge the very existence and foundations of Christendom. Although Muslims were initially a minority in the conquered territories, in time they became a majority, owing largely to mass conversions of local Christians.[16] In addition, those who remained Christian were Arabized, adopting Arabic language and culture. The response of Western Christendom was, with few exceptions, defensive and belligerent. Islam was a danger to be reckoned with. A seemingly impregnable Byzantine empire had buckled and risked being swept away during the seventh and eighth centuries. Muslim armies overran the Persian empire as well, conquered Syria, Iraq, and Egypt, and swept across North Africa and parts of southern Europe until they ruled most of Spain and the Mediterranean from Sicily to Anatolia. Ancient historical and theological affinities went unnoticed as the Christian West, Church and State, faced the onslaught of an enemy which it found easier to demonize and to dismiss as barbarian and infidel than to understand.

Non-Muslims in the Islamic State

The indigenous peoples in the conquered areas belonged to one of three major "scriptural" communities (People of the Book, *ahl al-kitab*):

Christian, Jewish, and Zoroastrian. For many non-Muslim populations in Byzantine and Persian territories already subjugated to foreign rulers, Islamic rule meant an exchange of rulers, the new ones often more flexible and tolerant, rather than a loss of independence. Many of these populations now enjoyed greater local autonomy and often paid lower taxes. The Arab lands lost by Byzantium exchanged Graeco-Roman rule for new Arab masters, fellow Semites with whom the populace had closer linguistic and cultural affinities. Religiously, Islam proved a more tolerant religion, providing greater religious freedom for Jews and indigenous Christians. Most of the local Christian churches had been persecuted as schismatics and heretics by a "foreign" Christian orthodoxy. For these reasons, some Jewish and Christian communities had actually aided the invading Muslim armies. Francis Peters has observed:

> The conquests destroyed little: what they did suppress were imperial rivalries and sectarian bloodletting among the newly subjected population. The Muslims tolerated Christianity but they disestablished it; henceforth Christian life and liturgy, its endowments, politics, and theology, would be a private not a public affair. By an exquisite irony, Islam reduced the status of Christians to that which the Christians had earlier thrust upon the Jews, with one difference. The reduction in Christian status was merely judicial; it was unaccompanied by either systematic persecution or blood lust, and generally, though not everywhere and at all times, unmarred by vexatious behavior.[17]

Just as Muslim rulers tended to leave the government institutions and bureaucracy intact, so too religious communities were free to practice their faith and be governed in their internal affairs by their religious laws and leaders. As previously mentioned, religious communities were required to pay a poll or head tax, in exchange for which they were entitled to peace and security; thus they were known as "protected people." The Islamic ideal was to fashion a world in which, under Muslim rule, idolatry and paganism would be eliminated, and all people of the book could live in a society guided and protected by Muslim power. While Islam was regarded as the final and perfect religion of God, others were to be invited, through persuasion first rather than the sword, to convert to Islam. Thus non-Muslims were offered three choices: (1) conversion to Islam and full membership in the community; (2) retention of one's faith and payment of a poll tax; or (3), if they refused Islam or "protected" status, warfare until Islamic rule was accepted.

The Crusades

Few events had a more shattering and long-lasting effect on Muslim–Christian relations than the Crusades. Two myths pervade Western perceptions of the Crusades: first, that Christendom triumphed; second, that the Crusades were simply fought for the liberation of Jerusalem. For many in the West, the specific facts regarding the Crusades are but dimly known.[18] Indeed, many do not know who started the Crusades, why they were fought, or how the battle was won. For Muslims, the memory of the Crusades lives on as the clearest example of militant Christianity, an earlier harbinger of the aggression and imperialism of the Christian West, a vivid reminder of Christianity's early hostility toward Islam. If many regard Islam as a religion of the sword, Muslims down through the ages have spoken of the West's Crusader mentality and ambitions. Therefore, for Muslim–Christian relations, it is less a case of what actually happened in the Crusades than how they are remembered.

The Crusades, which take their name from the "cross" (*crux* in Latin), were a series of eight military expeditions extending from the eleventh to the thirteenth centuries which pitted Christendom (the Christian armies of the Franks) against Islam (the Muslim armies of the Saracens). The eleventh century marked a turning point in the relationship of the West to the Islamic world.

> Up till 1000 the West was a poor, backward and illiterate region, precariously defending itself against the assaults of barbarous nations by land and sea. . . . All this while for four centuries, Islam enjoyed an internal peace and security, untroubled save for domestic wars, and thus was able to build up a brilliant and impressive urban culture. Now the situation was dramatically transformed. . . . Trade and commerce revived [in the West], towns and markets sprang up; the population increased . . . and the arts and sciences were cultivated on a scale unknown since the days of the Roman Empire.[19]

The West, emerging from the Dark Ages, mounted a counteroffensive to drive the Muslims out of Spain, Italy, Sicily, and the Mediterranean at a time when the Islamic world had experienced an upsurge in political and religious strife.

When his forces were decisively defeated by the Abbasid army in the late eleventh century, Byzantine Emperor Alexius I, fearing that Muslim armies would sweep across Asia and capture the imperial capital at Constantinople, appealed to the West. He called upon fellow Christian rulers and the Pope to turn back the Islamic tide by undertaking a "pilgrimage" to liberate Jerusalem and its environs from Muslim rule.

Jerusalem was a city sacred to all three Abrahamic faiths. It had been captured by Muslim armies in 638 during the period of Arab expansion and conquest. Under Muslim rule, Christian churches and populations were left unmolested. Christian shrines and relics had become popular pilgrimage sites for Christendom. Jews, long banned from living there by Christian rulers, were permitted to return, live, and worship in the city of Solomon and David. Muslims built a shrine, the Dome of the Rock, and a mosque, the al-Aqsa, near the Wailing Wall, the last remnant of Solomon's Temple, and thus a site especially significant to Judaism. Five centuries of peaceful coexistence were now shattered by a series of holy wars which pitted Christianity against Islam and left an enduring legacy of distrust and misunderstanding.

The Crusades were initiated by Pope Urban II's response to Emperor Alexius's plea. In 1095 Urban called for the liberation of the Holy Land from the infidel, appealing to an already established tradition of holy war. For the Pope, the call to the defense of the faith and Jerusalem provided an ideal opportunity to gain recognition for papal authority and its role in legitimating temporal rulers, and to reunite the Eastern (Greek) and Western (Latin) churches.

The Pope's battle cry "God wills it!" initially proved successful. The appeal to religion captured the popular mind and engaged the self-interest of many, producing a reinvigorated and relatively united Christendom. Christian rulers, knights, and merchants were driven by the political, military, and economic advantages that would result from the establishment of a Latin kingdom in the Middle East. Knights from France and other parts of Western Europe, moved by both religious zeal and hope of plunder, rallied and united against the "infidel" in a war whose ostensible goal was the liberation of the holy city: "God may indeed have wished it, but there is certainly no evidence that the Christians of Jerusalem did, or that anything extraordinary was occurring to pilgrims there to prompt such a response at that moment in history."[20]

The Crusades drew inspiration from two Christian institutions, pilgrimage and holy war: liberation of the holy places from Muslim rule partook of the character of both. Pilgrimage played an important role in Christian piety. Visiting sacred sites, venerating relics, and penance brought (its critics would say "bought") indulgences which promised the remission of sins. Jerusalem, central to the origins of Christian faith, was a symbol of the heavenly city of God and thus a major pilgrimage site. At the same time, the notion of holy war transformed and sacralized medieval warfare and its notions of honor and chivalry. Warriors were victorious whether they won their earthly battles or not. To rout the enemy meant honor and booty; the indulgences earned by all who fought in the Crusades guaranteed the remission of sins and entrance into paradise. To fall in battle was to die a martyr for the faith and gain immediate access to heaven despite past sins.

Caught off guard and divided, the initial Muslim response was ineffectual; the armies of the First Crusade reached Jerusalem and captured it in 1099. But Christian success was short-lived: "The Crusaders were . . . a nuisance rather than a serious menace to the Islamic world."[21] By the middle of the twelfth century, Muslim armies mounted an effective response. Under the able leadership of Saladin (Salah-al-Din, d. 1193), one of Islam's most celebrated rulers and generals, Jerusalem was reconquered in 1187. The tide had turned and the momentum would remain with Muslim forces. By the thirteenth century the Crusades had degenerated into intra-Christian wars, wars against enemies whom the papacy denounced as heretics and schismatics. Finally, the very fear that had initiated the Christian holy war, with its call for a united Christendom to turn back the Islamic tide, was realized in 1453 when the Byzantine capital, Constantinople, fell and, renamed Istanbul, became the seat of the Ottoman empire. A dream of Muslim rulers and armies originating in the seventh century had been fulfilled. Conversely, Christian fears and the continued threat of a powerful, expansive Islam now extended to Eastern Europe, much of which was brought under Ottoman rule.

The legacy of the Crusades depends upon where one stands in history. Christian and Muslim communities had competing visions and interests, and each one cherishes memories of its commitment to faith, and heroic stories of valor and chivalry against "the infidel." For many in the West, the assumption of a Christian victory is predicated on a romanticized history celebrating the valor of Crusaders, as well as a tendency to interpret history through the experience of the past two centuries of European colonialism and preeminent American power. Each faith sees the other as militant, somewhat barbaric and fanatical in its religious zeal, determined to conquer, convert, or eradicate the other, and thus an obstacle and threat to the realization of God's will. Their contention continued during the Ottoman period, through the next wave of European colonialism, and finally into the superpower rivalry of the twentieth century.

The Ottoman Empire: The Scourge of Europe

No sooner had the Crusades passed when Europe was once again faced with the power and might of the Muslim threat embodied in the Ottoman empire. The Ottoman empire was one of the three great imperial medieval Muslim sultanates: Ottoman, Safavid in Iran, and Mogul in India. As C. E. Bosworth has noted, more than any other empire since the early period of Arab conquest and expansion, "the Ottoman Turks struck terror into the hearts of Christian Europe, so that the Elizabethan historian of the Turks, Richard Knollys, described them as 'the present terror of the world.' "[22] Having taken Constantinople in 1453, the Ottomans proceeded to build a vast and extremely well-organized, hierarchical, and efficient state. The imperial capital at Istanbul, with a population that grew to 700,000—twice that of its European counterparts—became an international center of power and culture.[23] The Ottomans became the great warriors of Islam, creating a world empire that incorporated major Muslim centers like Cairo, Baghdad, Damascus, Mecca, and Medina. They threatened the heart of Europe for almost two centuries.

An unusual series of capable sultans (which included the celebrated Mehmet the Conqueror, 1451–81, and Suleiman the Magnificent, 1520–66), led Ottoman armies and navies that dominated much of the Mediterranean and the Indian Ocean. They created an empire whose size, prosperity, government, and culture rivaled that of the Abbasids. The Ottomans overpowered and subdued the Christian Balkan states, as well as much of the Middle East and North Africa. As in the early Arab conquests, their flexible policy toward Orthodox Christians and other religious minorities was often well received by the subject populations: "This live-and-let-live policy was in striking contrast to the fanatical bigotry of Christian states at the time. Balkan peasants in Mehmet's times used to say, 'Better the turban of the Turk than the tiara of the Pope.' "[24] But if many in the Balkans saw the Ottomans as liberators, the heartland in Europe seemed traumatized.

> Authors and clerics revived and refurbished the diatribes against the infidel which had characterized the period of the Crusades. . . . Cardinal Bessarion, writing to the Doge of Venice after the fall of Constantinople, set the tone for a century of abuse: "A city which was so flourishing . . . the splendour and glory of the East . . . the refuge of all good things, has been captured, despoiled, ravaged and completely sacked by the most inhuman barbarians . . . by the fiercest or wild beasts. . . . Much danger threatens Italy, not to mention the other lands, if the violent assaults of the most ferocious barbarians are not checked."[25]

Invective and propaganda were popularized by Bartholomew Gregevich of Croatia's best-selling work, *Miseries and Tribulations of the Christians Held in Tribute and Slavery by the Turks.*[26] The convenient, hostile stereotype and caricature of the Turks, informed more by passion than by reason, prevailed over the minority voices of those diplomats and scholars who had actually seen or dispassionately studied the Turks and, like the French political philosopher Jean Bodin, could observe:

> The King of the Turks, who rules over a great part of Europe, safeguards the rites of religion as well as any prince in this world. Yet, he constrains no one, but on the contrary permits everyone to live as his conscience dictates. What is more, even in his seraglio at Pera he permits the practice

of four diverse religions, that of the Jews, the Christian according to the
Roman rite, and according to the Greek rite, and that of Islam.[27]

The Ottoman threat contributed to the development of "Europe" as
the focus of a common identity and bond within a European Christendom torn apart by the Reformation, so that "Erasmus exhorted the
'nations of Europe'—no longer addressing them as the constituent powers of Christendom—to crusade against the Turks."[28]

Ottoman power and glory rested upon the development of a system for
training young men for military and administrative service. It produced a
first-class bureaucracy and military which relied heavily upon its religious
scholars (the *ulama*) and a corps of elite slave soldiers and officials, the
Janissaries. Young Christian males were taken from conquered populations of the Balkans, and later from Anatolia, converted to Islam, and
sent to special schools which trained and produced generations of Ottoman officials. The combination of extremely capable rulers and a well-
trained and disciplined military able to use gunpowder to advantage
enabled the Ottomans to conquer large areas of Arab and European territory: "It was the discipline and firepower of these troops (the Ottoman
army made use of artillery and hand-guns . . .) which did much to create
in Europe the image of Ottoman ferocity and invincibility."[29] Some eight
hundred years after the first Arab threat to Europe, Islam, now in the
hands of the Turks, seemed even more of a menace. Having subdued the
Balkans, they seemed poised to engulf Western Europe. From the fifteenth to the seventeenth century Ottoman forces seemed invincible to
European Christians. Yet the Ottoman naval defeat at Lepanto in 1571
was a turning point, hailed as a victory of Christian Europe over the Muslim Turks, and the successful defense of Vienna in 1683 confirmed the
decline of the Ottoman threat and the shift in power to a revitalized and
now self-confident Europe. The "Scourge of Christendom" would soon
become the "sick man of Europe."[30]

The Crusades and the Ottoman empire clearly show that, despite the
theological roots and affinities of Christianity and Islam, competing religious and political interests produced a history of confrontation and
warfare in the course of which Christian Europe for centuries often
found itself on the defensive against Muslim armies, seeming at times to
be fighting for its very existence.

European Images of Islam

The negative fallout from events in Christian–Muslim history is reflected in the view of Islam that emerges from Western literature and

thought. Although there were moments of contact, mutual knowledge, and constructive exchange, by and large the Muslim expansion into Europe, ranging from the Arab conquests through the Crusades and the Ottoman empire, produced alienation and distrust of Islam, which was primarily viewed as a threat to Christendom. This legacy, as Albert Hourani has noted, is "still present in the consciousness of Western Europe, still feared and still, in general, misunderstood."[31] Fear and disdain, coupled with European ethnocentrism, produced distorted images of Islam and Muslims and dissuaded scholars from serious study of Islam's contributions to Western thought. "[I]t was not until the years between the two World Wars that a serious effort was made to understand the contribution of Islam to the development of Western thought, and the effect on Western society of the neighborhood of Islam."[32]

As with the seventh-century Arab conquests, Christendom again experienced Islam as a double threat, both theological and political. The Crusades had for the first time made Islam all too familiar in medieval Europe, though not necessarily understood. R. W. Southern notes: "Before 1100 I have found only one mention of the name of Mahomet in medieval literature outside Spain and Southern Italy. But from the year 1120 everyone in the West had some picture of what Islam meant, and who Mahomet was. The picture was brilliantly clear, but it was not knowledge. . . . Its authors luxuriated in ignorance of triumphant imagination."[33]

This ignorance reflected not only lack of knowledge but also the all-too-common human tendency among educated and uneducated alike to denigrate and dehumanize the enemy, to assume a superior posture and dismiss that which challenges and threatens one's deepest beliefs or interests by labeling it inferior, heretical, fanatical, or irrational. Distorted portraits or caricatures of Muhammad and Islam were created—more accurately, fabricated—with little concern for accuracy. Often beliefs and practices such as polytheism, eating pork, drinking wine, and sexual promiscuity—which run directly counter to its most basic beliefs—were attributed to Islam and the Prophet. Muhammad was vilified as an impostor and anti-Christ who used magic and promiscuity to try to destroy the Church. As the non-Muslim author of an early biography of the Prophet produced in the West confessed, "It is safe to speak evil of one whose malignity exceeds whatever evil can be spoken."[34] The great epics of the times perpetuated ignorance and distortions, featuring idolatrous Muslims worshiping their chief god, Muhammad, "in synagogues (thereby bringing Islam closer to the equally unacceptable Jewish belief) or in 'mahomeries.' " Maxime Rodinson has observed: "Pure fiction, whose only object was to spur the reader's interest, was mixed in varying proportions with misrepresentations of belief which inflamed hatred of the foe."[35]

By the time of the Reformation, after centuries of fear and hostility, Islam proved a convenient tool in polemical attacks among Christians, a symbol of the dangers of the anti-Christ. Martin Luther saw Islam "in the medieval way, as a movement of violence in the service of the anti-Christ; it cannot be converted because it is closed to reason; it can only be resisted by the sword, and even then with difficulty."[36]

In later centuries Islam continued to be used as a foil for authors who championed Enlightenment principles and virtues. Voltaire's *Fanaticism, or Muhammad the Prophet* portrayed the Prophet as a theocratic tyrant, and Ernest Renan, in an oft-cited lecture, championed science, reason, and human progress by dismissing Islam as incompatible with science, and the Muslim as "incapable of learning anything or of opening himself to a new idea."[37] These stereotypes of a static, irrational, retrogressive, antimodern religious tradition were to be perpetuated by scholars and development theory in the twentieth century.

Though the Islamic and Christian worlds take enormous pride in their faith and rich traditions of learning and civility, the historical dynamics of Christian–Muslim relations often found the two communities in competition, and locked at times in deadly combat, for power, land, and souls. As a result they were often enemies, rather than People of the Book striving in a common quest to obey and serve their Lord. For Christendom, Islam proved a double threat, religious and political, which often threatened to overrun Europe, first at Poitiers and finally at the gates of Vienna. It was not in jest that some historians noted that had Muslim armies not been turned back at Poitiers, the language of Oxford, as indeed of Europe itself, might have been Arabic! A Christian Church convinced of its possession of truth and its preordained mission to save the world legitimated papal and imperial designs. Moreover, it fostered a sense of superiority and righteousness which provided a rationale for the denigration of the infidel religiously, intellectually, and culturally. These same attitudes made the successes of Muslim armies and the rapid spread of Islam by soldiers, traders, and missionaries that much more of a challenge to Christian faith and power. If the first ten centuries seemed a lopsided contest in which Christendom was more often than not literally or figuratively under siege, the dawn of European colonialism signaled a shift in power: thereafter, colonialism would dominate the history and psyche of Muslims, and continues seriously and at times dramatically to affect relations between Islam and the West today. As the Iranian Revolution of 1978–79 and, more recently, the Gulf War of 1991 have revealed, images of Christian Crusaders and Western imperialism remain a living legacy, an experience very much alive in Muslim consciousness and political rhetoric.

3

The West Triumphant: Muslim Responses

Two themes dominate the first half of twentieth-century Muslim history: European imperialism, and the struggle for independence from colonial rule. Few events were more far-reaching and influential in the relationship of Islam to the West than the experience of European colonialism. The theme of European colonialism and imperialism, their impact in the past and their continued legacy, remains alive in Middle East politics and throughout the Muslim world from North Africa to Southeast Asia. Intertwined with colonial rule was the emergence of nationalist movements in the decades-long battle for independence. In the post-independence period political elites in emerging Muslim states turned to liberal nationalism and Arab socialism in nation building. These experiments failed, damaging the credibility of nationalism and contributing to the Islamic resurgence. Issues of foreign domination and dependence remain a bitter memory as well as a continued threat in the eyes of many Muslims today.

Colonialism quite literally altered the geographic and institutional map of the Middle East, or perhaps more accurately, it often drew the boundaries and appointed the leaders for much of the modern Muslim world. It replaced or transformed indigenous political, social, economic, legal, and educational institutions and explicitly as well as implicitly challenged Muslim faith and culture. Countries that welcomed European trade missions in the sixteenth and seventeenth centuries found themselves transformed by the nineteenth century into European colonies or protectorates.

The image of Islam as both a potential threat to the Christian West and a retrogressive force and thus a source of Muslim backwardness and decline dominated the worldview of European colonialism. It provided a ready-made rationale for "crown and cross." Colonial officials and Christian missionaries became the footsoldiers of Europe's expansion and imperial hegemony in the Muslim world. The British spoke of the "white man's burden" and the French of their "mission to civilize." As the balance of power and leadership shifted from the Muslim world to Europe, modernity was seen as the result not simply of conditions that produced the Enlightenment and the industrial revolution, but also of Christianity's inherent superiority as a religion and culture.

> This was Civilization, as proud, imperial-trivial nineteenth-century Europe called it, demeaning everything non-European with an insensitivity. . . . (Was it triumphalism, lack of experience, a parvenu spirit, and too abrupt an entrance into the wider field of human civilization?) . . . Europe did not become aware of how original its civilization was until it compared it with that of others. . . . Then the internal disparities faded from view and the white man affirmed himself and closed ranks in solidarity against everything outside.[1]

Lord Cromer, British consul in Cairo from 1883 to 1907 during Britain's occupation of Egypt, epitomized this outlook. Albert Hourani has summarized Cromer's attitude as a "benevolent" imperialism. Islam is seen as a

> "noble monotheism," but as a social system it "has been a complete failure": Islam keeps women in a position of inferiority, it "crystallizes religion and law into an inseparable and immutable whole, with a result that all elasticity is taken out of the social system"; it permits slavery; its general tendency is towards intolerance of other faiths; it does not encourage the development of the power of logical thought. Thus, Muslims can scarcely hope to rule themselves or reform their societies; and yet Islam can generate a mass feeling which, in a moment, can break whatever brittle bonds the European reformer has been able to establish with those he is trying to help. The fear of the "revolt of Islam" is never far from Cromer's thoughts.[2]

European colonialism posed both a political and a religious challenge. It abruptly reversed a pattern of self-rule in the Muslim world which had existed from the time of the Prophet. By and large, the vast majority of the Muslim community had possessed a sense of history in which Islam had remained triumphant. Despite past divisions, civil wars, and revolts as well as invasion and occupation, Islam had prevailed—Muslims had ruled Muslims. To be a Muslim was to live in a state which at least nominally was a Muslim community guided by the laws and institutions of Islam. Even the seemingly catastrophic end of the Abbasid Caliphate with the fall of Baghdad to the Mongols in 1258 was followed by the conversion of the Mongol conquerors and the continuance of an Islamic world order composed of Muslim sultanates which extended from Africa to Southeast Asia. However, this sense of Muslim history and belief now seemed to be unraveling, owing to internal as well as external threats to the identity and fabric of Islamic society. Muslim communities had already been struggling with internal problems. The external challenge and threat of subjugation by Europe came on the heels of a powerful wave of Muslim religious revivalism in the eighteenth century which had addressed the internal sociomoral decline of the community. The political challenge of European colonialism was intensified by the threat posed by the wave of Christian missionary activity which sought to win souls for Christ and openly questioned the viability of Islam in the modern world.

> The degraded state of the Muslim world made it an obvious target for Christian missionaries. The proselytizing crusade was launched with renewed vigor and quickly spread. . . . In keeping with the common beliefs of their time and normal human inclinations, the missionaries credited the triumphs of European nations to Christianity while blaming the misfortunes of the Muslim world on Islam. The perception was that, if Christianity was inherently favorable to progress, then Islam must, by its nature, encourage cultural and developmental stagnation.[3]

The external threat to Muslim identity and autonomy intensified profound religious as well as political questions for many in the Muslim world. What had gone wrong? Why had Muslim fortunes been so thoroughly reversed? Was it Muslims who had failed Islam or Islam that had failed Muslims? How were Muslims to respond?

Islamic Renewal and Reform

Although modern Islamic reform is often simply presented as a response to the challenge of the West, in fact its roots are both Islamic (its revivalist tradition) and Western (a response to European colonialism).

Islam possesses a rich, long tradition of Islamic revival (*tajdid*) and reform (*islah*).[4] Down through the ages, individuals (theologians, legal scholars, Sufi masters, and charismatic preachers) and organizations undertook the renewal of the community in times of weakness and decline, responding to the apparent gap between the Islamic ideal and the realities of Muslim life. As with all things, a return to Islam—that is, to the fundamentals: the Quran, the life of the Prophet, and the early Islamic community—offered the model for Islamic reform.

During the eighteenth and nineteenth centuries, revivalist leaders and movements had sprung up across the Islamic world: the Mahdi (1848–85) in the Sudan, the Sanusi (1787–1859) in Libya, the Wahhabi (1703–92) in Saudi Arabia, the Fulani in Nigeria (1754–1817), the Faraidiyyah of Hajji Shariat Allah (1764–1840) in Bengal, the militant movement of Ahmad Brelwi (1786–1831) in India, and the Padri in Indonesia (1803–37).[5]

Most revivalist movements were primarily internally motivated; they responded to a decline whose root cause was identified as being within the Islamic world. The powerful revivalist spirit that gripped the Islamic world during the eighteenth century was a response to economic and sociomoral decline, military defeats, and political divisions within the imperial sultanates (Ottoman, Safavid, and Mogul) and beyond. Despite differences, all were movements whose goal was the moral reconstruction of society. They diagnosed their societies as being internally weak and in decline politically, economically, and religiously. The cause was identified as Muslim departure from true Islamic values brought about by the infiltration and assimilation of local, indigenous, un-Islamic beliefs and practices. The prescribed cure was purification through a return to "true Islam." Each movement resulted in the formation of organizations, a society of "true believers" within the broader society, that combined religious commitment with militant political activism in order to purify Muslim communities. The process of Islamic renewal and reform was based upon a return to the fundamental sources of Islam. In time, several of these movements led to the creation of new states: the Mahdi in the Sudan, the Sanusi in Libya, the Fulani in Nigeria, and the Wahhabi in Saudi Arabia.

Emulating the example of the prophet Muhammad, revivalist movements transformed their societies through a religiously legitimated and inspired sociopolitical movement. The ideological worldview of revivalist movements had an impact not only on their societies but also on Islamic politics in the twentieth century. The key ideological components of their program were: (1) Islam was the solution; (2) a return to the Quran and the Sunnah (model, example) of the Prophet was the method; (3) a community governed by God's revealed law, the Sharia, was the goal; and (4) all who resisted, Muslim or non-Muslim, were enemies of God. Members of the community, like the early Muslims of the seventh century, were trained in piety and military skills as these movements spread God's rule through preaching and jihad.

European Colonialism

By the nineteenth century a clear shift of power had occurred, as the decline of Muslim fortunes reversed the relationship of Islam to the West. Increasingly, Muslims found themselves on the defensive in the face of European expansion. Whereas the primary challenge to Islamic identity and unity in the eighteenth and nineteenth centuries was generally seen as internal, the real threat of the West was not experienced until the late nineteenth and early twentieth century. It constituted a singular challenge to Islam politically, economically, morally, and culturally. European colonialism and imperialism threatened Muslim political and religiocultural identity and history. The impact of Western rule and modernization raised new questions and challenged time-honored beliefs and practices. With the dawn of European domination of the Muslim world, the image, if not always the reality, of Islam as an expansive worldwide force had been shattered.

The map of the Muslim world after World War I revealed the extent of foreign dominance: the French in North, West, and equatorial Africa and the Levant (Lebanon and Syria); the British in Palestine, Transjordan, Iraq, the Arabian Gulf, and the Indian subcontinent; and in Southeast Asia, the British in Malaya, Singapore, and Brunei, and the Dutch in Indonesia. Where Muslims retained power, in Turkey and Iran, they were constantly on the defensive against the political and economic ambitions of the British, French, and Russians, whose inroads and machinations threatened their independence and stability. Given the overwhelming presence and political, military, and economic superiority of Europe, it is not surprising that, as John Voll has observed, "[t]he old-fashioned Western imperialist [Lord Cromer] could complacently conclude that Islam's 'gradual decay cannot be arrested by any modern palliatives however skillfully they are applied.' "[6]

Muslim views of the West and responses to its power and ideas varied from rejection and confrontation to admiration and imitation. However, the prevailing mood was one of conflict and competition. For many, colonialism conjured up memories of the Crusades; the European challenge and aggression was but another phase of militant Christianity's war with Islam; Europe was the enemy that threatened both the faith of Islam and the political life of the Muslim community. The political crisis precipitated by European colonialism was accompanied by a spiritual one: "The fundamental spiritual crisis in Islam in the twentieth century stems from an awareness that something is awry between the religion which God has appointed and the historical development of the world which He controls."[7]

Muslim images of a Crusader West were reinforced by the policies of colonial powers. Colonialism was experienced as a threat to Muslim identity and faith. Implicit in its policies and explicit in the statements of many government officials and missionaries was the belief that Europe's expansion and domination were due to its inherent Christian cultural superiority. Educating the "natives" in the language, history, and sciences of the West and Christian virtues was part of an "enlightened" policy to civilize. One did not have to look far for statements which substantiated the worst Muslim fears. As a result, the struggle against European colonialism often appropriated the rhetoric of a war between Christendom and Islam. While French apologists spoke of the "battle of the Cross against the Crescent," a defeated Algerian Muslim officer declared, "We have fought to this day in defense of our liberties and religion."[8] The master–servant relationship inherent in a good deal of European colonial rule generated mutual stereotypes and distrust that reflected an adversarial relationship:

> The Egyptian, as viewed by the British, was a portly parody of a petty French official, a boastful coward, a turbaned Muslim fanatic, a noisy agitator blind to the benefits British rule had given the country, or a "wog" (for "wily Oriental gentleman") selling dirty postcards in the bazaar. The Egyptians saw the British as coldhearted, exclusive (it was a sore point that, for many years, the only Egyptians that could enter Cairo's posh Gezira Sporting Club were servants), mercenary, and power-mad.[9]

Europe came not only with its armies of bureaucrats and soldiers but also with its Christian missionaries. The double threat of colonialism was that of the crown and the cross. The mutual relationship between the clergy and the government and military was proclaimed by France's Marshal Bugeaud, who praised their "grands rapports," commenting that the clergy "gain for us the hearts of the Arabs whom we have subjected to force of arms."[10] The preachers and missionary institutions (churches, schools, hospitals, and publishing houses) were regarded by many Muslims as an arm of imperialism, one aspect of a policy that displaced indigenous institutions, supplanted local languages and history with Western curricula, and seduced souls through schools and social welfare. The French seizure of the Grand Mosque of Algiers and its conversion into the cathedral of Saint-Philippe, with the French flag and the cross on its minaret, symbolized the threat of Christianity. France's patronizing disdain was boldly articulated by the archbishop of Algiers when he stated that the Church's mission was to convert Arab Muslims from "the vices of their original religion generative of sloth, divorce, polygamy, theft, agrarian communism, fanaticism, and even cannibalism."[11]

Muslim Responses to the Challenge of Colonialism

Four diverse Muslim responses to the West took shape: rejection; withdrawal; secularism and Westernization; and Islamic modernism.

Rejection and Withdrawal

For many Muslims, the Prophet's response to his Meccan detractors was the answer—emigration (*hijra*) or sacred struggle (jihad): leaving a territory no longer under Muslim rule, or fighting to defend and defeat the infidel. Although Christians had always been regarded as believers, People of the Book, European Christian colonizers were now rejected as infidels, the enemies of Islam. While resistance or confrontation initially proved attractive, emigration proved impractical for large numbers of people and, given the superior military strength of Europe, holy war was doomed to defeat. For many religious leaders, the alternative was simply to refuse to deal with their colonial masters, to shun their company, schools, and institutions. Any form of cooperation was regarded as capitulation to the enemy or treason. Modern European education was condemned as alien, superfluous, and a threat to religious belief. Muslims did not have to look far to find Europeans whose statements reinforced their fears; said one English author: "The luxuriant religions of Asia shrivel into dry sticks when brought into contact with the icy realities of Western sciences."[12]

Secularism and Westernization

However, Muslim responses varied. If some preached rejection and resistance, others were eager to learn from and emulate the strength of Europe, to modernize. The conclusion of one scholar regarding Muslim reactions in the Indian subcontinent is generally true for much of the Muslim world: "Muslim reaction to English education was by no means uniform . . . the Muslim reaction to modernity varied from blind and self-defeating hostility to reasonable cooperation with British educational policy."[13] For many, the realities of European ascendancy had to be acknowledged and dealt with, and its lessons discerned in order to survive.

Muslim rulers in the Ottoman empire, Egypt, and Iran looked to the West to develop military, economic, and political modernization programs based upon European learning and technology. They sought to emulate the strength of the West, to develop a modern trained and equipped military and bureaucracy, and to appropriate the science that provided modern weapons. European teachers and schools were imported. Educational missions were sent to Europe, where Muslims studied languages, science, and politics. Translation bureaus and printing presses were established to translate and publish Western works. A new intellectual elite was born—modern, educated, and Western-oriented. However, change through emulation came from above, initiated and imposed by rulers, in reaction to the external threat of European expansionism and not as a response to internal societal pressures. The state initiated this change, and a small elite implemented and were the primary beneficiaries of reform. Though Islamic rationales were employed by some to legitimate the transformation, implicit in this process was the gradual acceptance of a secular outlook that restricted religion to personal life while turning to the West for development models in public life. The result was a series of military, administrative, educational, economic, legal, and social reforms, strongly influenced and inspired by the West, to "modernize" Muslim societies. The traditional Islamic basis and legitimacy of Muslim societies were slowly altered as the ideology, law, and institutions of the state, indebted to imported models from the West, were increasingly secularized.

The modernization through Western models initiated by Muslim rulers was primarily motivated by a desire to strengthen and centralize their power, not to share it. Political participation was not a government priority. The rulers' primary interest was in military, bureaucratic, and technological reform, not in substantive political change. Thus for example, though the Ottoman sultan Abdulhamid II promulgated the First Ottoman Constitution in 1876, and the Shah in Iran capitulated to pressure and established a National Consultative Assembly in 1906, in both cases attempts to bring about serious constitutional reforms limiting the absolute power of the rulers were thwarted or crushed.

A major result of modernization was the emergence of new elites and a growing bifurcation of Muslim society, epitomized in its legal and educational systems. The coexistence of traditional religious and modern secular schools, each with its own curriculum, teachers, and constituencies, produced two classes with divergent worldviews: a modern Westernized elite minority and a more traditional, Islamically oriented majority. The process also eroded the traditional bases of power and authority of religious leaders, as new classes of modern trained elites assumed positions of

importance in government, education, and law, positions which had always been the province of the *ulama.*

Islamic Modernism

A fourth response to the challenge of the West, the Islamic modernist
movement, sought to bridge the gap between Islamic traditionalists and
secular reformers. Islamic modernists incorporated the internal community concerns of eighteenth-century revivalism with the need to respond
to the threat of European colonialism and the demands of modernity.[14]
Like secular reformers, Islamic reformers responded to European colonialism and were influenced by their perception of the "Success of the
West." The West was strong and successful; Muslims were weak and
subject to domination and dependence. Thus they believed that the
sources of the West's strength must be accommodated and assimilated.
In the latter half of the twentieth century, this posture would stand in
sharp contrast to Islamic activism (fundamentalism), with its denunciation of the "neocolonialism of the superpowers" and its emphasis on the
"Failure of the West" and the autonomy and self-sufficiency of Islam.[15]

Islamic modernism, like much of the modern Muslim response to the
West in the twentieth century, had an ambivalent attitude toward the
West, a simultaneous attraction and repulsion. Europe was admired for
its strength, technology, and political ideals of freedom, justice, and
equality, but often rejected for its imperialist goals and policies. Reformers like Jamal al-Din al-Afghani and Muhammad Abduh in Egypt, Morocco's Allal al-Fasi, Tunisia's Abd al-Aziz al-Thalabi, Algeria's Abd al-
Hamid Ibn Badis, and Sayyid Ahmad Khan and Muhammad Iqbal in
the Indian subcontinent, argued the compatibility of Islam with modern
science and the best of Western thought. They preached the need and
acceptability of a selective synthesis of Islam and modern Western
thought; condemned unquestioned veneration and imitation of the past;
reasserted their right to reinterpret (*ijtihad*) Islam in light of modern
conditions; and sought to provide an Islamically based rationale for
educational, legal, and social reform to revitalize a dormant and impotent Muslim community. In contrast to eighteenth-century revivalist
movements, Islamic modernism did not seek to restore a pristine past
but instead wished to reformulate its Islamic heritage in response to the
political, scientific, and cultural challenge of the West. It provided an
Islamic rationale for accepting modern ideas and institutions, whether
scientific, technological, or political (constitutionalism and representative government). For most of these reformers, the renaissance of the
Muslim community was the first step to national independence or libera-

tion from the hated yoke of colonialism—the restoration of Muslim power. Muslims, they believed, must look to Islam, their source of strength and unity, but learn the secrets of Western power in order to cast off foreign rule and regain their identity and autonomy.

Jamal al-Din al-Afghani (1838–97) epitomized the concerns and program of Islamic modernism. Born and educated in Iran and then in British India, Afghani was a teacher and tireless political activist who roamed the Muslim world from Egypt to India, calling upon Muslims to arise from their lethargy and reclaim their God-ordained purpose and identity. While he taught in Egypt, served as an adviser to the Shah of Iran, and traveled in Europe, he preached a message that challenged both Muslim and European authorities, citing "the danger of European intervention, the need for national unity to resist it, the need for a broader unity of the Islamic peoples, the need for a constitution to limit the ruler's power."[16] These themes of anti-imperialism, Arab unity, Pan-Islam (Islamic solidarity), and constitutionalism were a major part of his legacy.

Afghani proved a thorn in the side of Muslim and British rulers alike. He and his disciples were involved in the Urabi nationalist revolt against British and French influence in Egypt in 1882. Afghani was subsequently deported from Iran in 1891 for his instigation and support of the Tobacco Protest. Later, after being invited to Istanbul by the Ottoman sultan Abdulhamid, Afghani was placed under virtual house arrest in 1896 when Iran's Nasir al-Din Shah was assassinated by a reputed follower of Afghani's.

Afghani blamed the ills of the Muslim world, its political decline and religious stagnation, on European expansionism, autocratic rulers, and the religious establishment. Critical of the *ulama* and their retrogressive interpretation of Islam, Afghani preached a message of renewal and change. Formulated in response to Europe and its criteria for modernity, it sounded very much like a "Protestant Islam." Indeed, Afghani believed that Islam, like Christianity before it, needed a Martin Luther and a reformation.[17] Islam was the religion of progress and change, of reason and science, a religion with a strong work ethic. Reformers argued that these characteristics were integral to Muslim history, the source of Islam's past strength and glory. Muslims had produced and possessed a rich Islamic civilization; thus, Afghani argued, reason, philosophy, and science were not foreign to Islam, were not simply products of the West. Similarly, Islam had provided the social bond which unified and guided a once triumphant community. Muslim unity, like anti-imperialism, remained a prerequisite for political and cultural independence.

For Afghani the revitalization of a subjugated community could not be achieved by ignoring or rejecting the West but by active engagement and confrontation. The West was both the problem and part of the solution. On the one hand, Europe had subdued and threatened the identity and autonomy of the community. On the other hand, the Islamic community nationally and transnationally must learn from the West, identify and harness the sources of its power. Thus in addition to science and technology, Afghani appropriated political ideas such as constitutionalism and political participation through elected assemblies. The revitalization of Islam and Muslim solidarity were the keys to attain the ultimate goal, independence from the West and the restoration of Muslim fortunes.

If Afghani was primarily a political activist, his protegé Muhammad Abduh (1849–1905) was the developer of the intellectual and social reformist dimensions of Islamic modernism. Abduh's Salafiyya movement sought legitimacy through identification of its Islamic modernist reformism with the elders (*salafi*) of the early Muslim community, those who followed the example of the Prophet. Abduh was an Egyptian trained as a religious scholar, a member of the *ulama*. He taught at al-Azhar University, the oldest center of Islamic learning, and at Dar al-Ulum, a new college that incorporated a modern curriculum to prepare Azhar graduates for government positions. The most enthusiastic of Afghani's early students, Abduh worked closely with Afghani, publishing articles on sociopolitical reform, and was exiled to Paris with Afghani for his participation in the Urabi revolt. In France they formed a secret society and published a newspaper that continued to preach Afghani's message of Islamic reform and anticolonialism.

After his return from exile, Abduh turned away from political activism and focused on intellectual, religious, educational, and social reform. His published works included the journal *al-Manar* (*The Beacon* or *Lighthouse*), which treated Quranic exegesis and theology. He preached the compatibility of revelation and reason, condemned the blind following of tradition (*taqlid*), championed the legitimacy of and need for a reinterpretation of Islam to respond to the demands of modern life. Abduh worked to reform the *ulama,* in particular the curriculum of al-Azhar University, and the religious courts. He provided a rationale for the reform of Islamic law, arguing that while laws concerned with worship of God were immutable, Islam's social legislation was capable of substantive change. As mufti (chief religious leader) of Egypt, Abduh's official opinions covered a broad range of reforms from the permissibility of European attire to banking interest, marriage, and divorce.

Muhammad Abduh was also an early champion of legal and educational reforms to improve the status of Muslim women. A critic of polygamy and its negative effect on the Muslim family, he argued that it had been permitted as a concession to prevailing social conditions in Arabia at the time of the Prophet. Abduh offered a modernist interpretation of the Quran which concluded that the Quranic ideal (Quran 4:3 and 4:129) was in fact monogamy, since the Quran's permission for more than one wife was contingent upon equal justice and impartiality, both of which were a practical impossibility. Muhammad Abduh's Quranic interpretation was adopted by most Islamic reformers and provided the modernist rationale for many governments' Muslim family-law reforms restricting polygamy.

However, it was Abduh's associate Qasim Amin (1863–1908), a lawyer and judge, who developed the feminist dimension of Islamic modernism. In two controversial books (*The Emancipation of Women* and *The New Woman*) he denounced the subjugation of Muslim women as un-Islamic and a major cause for the deterioration of the family and society. Amin linked the emancipation of women to the nationalist cause. The bondage of women retarded national development. He argued the equality of the sexes in Islam, denounced veiling and social seclusion as un-Islamic, criticized the ills of arranged marriages, and a male's unfettered right to divorce and a wife's inability to do so. Although criticized and attacked by many religious and nationalist leaders who accused him of aping the West, Amin became an inspiration to Egyptian feminists like Huda Shaarawi a generation later.

While Islamic modernists like Jamal al-Din al-Afghani in the Middle East and Sir Sayyid Ahmad Khan (1817–98) in the Indian subcontinent shared a modern reformist agenda, they differed in their political orientation and goals, in particular their attitude toward the West. In contrast to Afghani, whose anticolonialist ideas and actions were so influential in the Middle East, Ahmad Khan was a British loyalist. Ahmad Khan, who had lived through the Mutiny of 1857 (which nationalists would later call the first war of independence), committed his life to the survival of the Muslim community in British India. He had two goals: to restore a debilitated and defeated Muslim community, and to reassure the British that Islam itself was not a threat to British interests. He called for a new theology to respond to modern change. His writings and activities contributed to this process of reinterpretation and reform. Ahmad Khan concentrated specifically on producing a new generation of Muslim leaders through educational reforms. He created the Anglo-Muhammadan Oriental College (in 1920 renamed the Aligarh Muslim University), modeled on Cambridge University. At the same time Ahmad Khan sought to counter both the *ulama* who dismissed the British as "enemies of Islam," and the British who regarded the Muslims as an inherent political threat because of their allegiance to Islam. Thus he opposed the Muslim tendency to advocate a policy of withdrawal and, contrary to prevailing British belief, argued that Muslims could be loyal citizens in a non-Muslim state. Despite his accomplishments, his close identification with the West condemned him in the eyes of those who chafed under colonial rule. Ahmad Khan's strong affinity for the West, symbolized by his decision to wear Western clothing, adopt a European life-style, and accept knighthood by Queen Victoria, brought strong criticism from many *ulama* and anticolonialists who dismissed his loyalism and reformism as political and cultural capitulation.

Muhammad Iqbal (1875–1938) in India and Taha Husayn (1889–1973) in Egypt represented the next stage in modernist reform, in which nationalist movements grew and a more secular orientation emerged. They exemplify the dual winds of change: Islamic modernist and secular nationalist.

Iqbal combined a traditional upbringing with an intimate knowledge of the West. He studied in Germany and England, earning a doctorate in philosophy at Munich and a law degree at London. Though a lawyer by profession, he is remembered as the great poet–philosopher of southern Asia. Iqbal combined what he thought to be the best of the East and the West, his Islamic heritage and Western philosophy (Hegel, Bergson, Fichte, and Nietzsche), to produce his own synthesis and reinterpretation of Islam. The reformist thrust of his thinking and its openness to Western thought was summarized in *The Reconstruction of Religious Thought in Islam:*

> No wonder then that the younger generation . . . demand a fresh orientation of their faith. With the reawakening of Islam, therefore, it is necessary to examine, in an independent spirit, what Europe has thought and how far the conclusions reached by her can help us in the revision, and if necessary reconstruction of theological thought in Islam.[18]

Like Afghani, Iqbal spoke of the need for an Islamic reformation like that which Christianity had undergone: "We are today passing through a period similar to that of the Protestant Reformation in Europe and the lesson which the rise and outcome of Luther's movement teaches should not be lost on us."[19]

Iqbal borrowed from the West but was not uncritical of it. He leveled sharp criticism at European colonialism and imperialism. Though an admirer of the accomplishments of the West, its dynamic spirit, intellectual tradition, and technology, he denounced the excesses of colonialism and imperialism, the exploitation of capitalism, the atheism of Marxism, and the moral bankruptcy of secularism. Like Ahmad Khan, Iqbal argued that, contrary to the Western caricature of Islam as a "religion of holy war," Islam was a religion of peace: "All forms of political and social disturbance are condemned . . . the ideal of Islam is to secure social peace at any price."[20] In contrast to Ahmad Khan and more like Afghani, Iqbal wanted to resuscitate the Muslim community so that it could reclaim its political independence and rightful place in history. Iqbal attempted to develop alternative Islamic models for modern Muslim societies. Drawing on Islamic traditions, he sought to "rediscover" Islamic principles and values that would provide the basis for Islamic versions of Western concepts and institutions such as democracy and parliamentary government.

Although Iqbal believed that nationalism was antithetical to the Islamic ideal of a transnational community, or Pan-Islam, he nevertheless accepted its practical necessity and utility. Muslims must gain independence and rebuild their local and regional communities. In contrast to Sir Sayyid Ahmad Khan's loyalism, Iqbal was an early voice both for independence from Britain and for Muslim nationalism in the subcontinent. Though originally an Indian nationalist, his concern for the identity and welfare of a Muslim minority in a Hindu-dominated state led Iqbal to turn from the dream of a united India, join the Muslim League, and call for a separate Muslim state.

In contrast to Iqbal's Islamic reformism, Taha Husayn, a former student of Muhammad Abduh, exemplifies the secular European drift of a new generation of Egyptian youth. His life and upbringing reflect the bifurcation of education and the split identity in Egyptian society. Although blind from an early age, he attended a village religious school, then was educated at Cairo's two premier institutions of higher learning, al-Azhar University, the famed Islamic center of learning, and the Egyptian (Cairo) University, the new modern national university. This was followed by four years (1915–19) of study in France, where he earned a doctorate from the Sorbonne. Husayn became an internationally known writer and educator and served as minister of education (1950–52).

Taha Husayn represents that group of Abduh's disciples who became leaders in politics and intellectual life but believed that the modern needs of society would best be served by the separation of religion and politics. Perhaps no book better represents the cultural crossroads that Muslims faced in the early twentieth century, and the Western orientation of many emerging Muslim elites, than Husayn's *Future of Culture in Egypt*. It embodies the rationale of those who concluded that future strength was best achieved not by a return to an Islamic past or the path of Islamic modernism, but rather by aggressive pursuit of Western-oriented liberal, secular reform. Although he was careful to speak of selective borrowing, the enormous pull of the West as a model for success was obvious in both his discourse and the degree of his reliance on Western models of development.

Taha Husayn and many of his generation were not critics of the West, but its unabashed admirers. Though he fought for political independence from Europe, he maintained that the roots of Egypt (as well as its future) and the roots of Islam were inextricably bound to the West. Indeed, he argued that Egypt was not of the East but of the West! He stressed their shared common religious and geographic origins (the one God, the Mediterranean). Just as Europe did not become Western when it embraced Christianity, whose roots were in the Middle East, so too Egypt did not become Eastern when it hastened to accept Islam. Similarly, Husayn maintained that the "essence and source of Islam are the essence and source of Christianity. . . . No, there are no intellectual or cultural differences to be found among the peoples who grew up around the Mediterranean and were influenced by it."[21] Husayn asserted that secularism had long been part of Egypt's tradition: "From earliest times Muslims have been well aware of the now universally acknowledged principle that a political system and a religion are different things, that a constitution and a state rest, above everything else, on practical foundations."[22] Thus, he concluded, it was only natural that modern Egyptians should adopt European customs and institutions from table manners and railroads to political, legal, and educational systems.[23]

In many ways, the secular drift of Taha Husayn (and other former disciples of Afghani and Abduh) is symptomatic of the mixed legacy of Islamic modernism. It was both a success and a failure. Modernists did provide Islamically based rationales for modern reform. They did offer an alternative to the rejectionist tendencies of many religious leaders and the uncritical assimilationist proclivities of secularists. Through the modernists' writings, publications, teaching, and the establishment of educational and social institutions, Islamic modernist ideas and values became part of Muslim discourse and would in time become part of mainstream Muslim thought. Afghani and Abduh's legacy influenced not only the Arab heartland and North Africa but also distant Indonesia. Ahmad Khan and Iqbal were major cultural and intellectual forces throughout southern Asia and beyond. However, modernist leaders themselves failed to produce organizations that would systematically develop and implement their ideas. Nowhere is this more evident than in the tendency of many of their disciples to turn toward a more secular path. Taha Husayn is a prime example of this tendency, as is Saad Zaghlul, the great Egyptian nationalist leader.

Islamic modernism was primarily an intellectual movement. While it did not produce a unified movement or enduring organizations, its legacy was substantial in its influence on the Muslim community's development and its attitude toward the West: Islamic modernism reawakened Muslims to a sense of past power and glory; reinterpreted and produced a modern ideological interpretation of Islam; and demonstrated the compatibility of Islam with modern Western sociopolitical reform. Most reformers distinguished between adopting Western ideas and technology and rejecting Western imperialism; indeed, they promoted the ideas of anticolonialism and Muslim unity, autonomy, and independence. For these reasons, men like Jamal al-Din al-Afghani and Muhammad Iqbal came to be remembered as fathers of Muslim nationalism.

Nationalist Movements and the Long March to Independence

The period between the two world wars was dominated by two interrelated issues, national identity (nationalism) and independence. Ironically, secular and Islamic reformers saw the West as being at one and the same time a positive and negative force. On the one hand, nationalism was a reaction to Western imperialism, to European colonial rule. On the other hand, it was in a sense also the product of a century of Westernizing reform. Many of those who led nationalist and independence movements owed their training to the West and were influenced by the liberal nationalist beliefs and ideals of the French Revolution (liberty, equality, fraternity) and, more specifically, modern Western political values and institutions such as democracy, constitutional government, parliamentary rule, individual rights, and nationalism. In contrast to the traditional Islamic ideal in which political loyalty and solidarity rest in a transnational Islamic community (*ummah*) based upon common belief, modern nationalism represented the notion of national communities based not upon religion but upon common language, territory, ethnic ties, and history.

Islam played an important role in the development of anticolonial independence movements and modern nationalism. It proved a factor to varying degrees in the development of local and regional nationalisms: Arab, Egyptian, Algerian, Tunisian, Moroccan, Iranian, Pakistani, Malaysian, and Indonesian. However, if religion was one factor, it was far from the only one. The appeal to Islam varied regionally and from country to country, conditioned by local contexts. In some areas Islam was a prominent ingredient in nationalism, while in others it was subordinated to secular nationalism. In many parts of the Muslim world, Islamic reformism (the modernist movement) and nationalism joined together to form a potent force. The major themes of Islamic reformism inspired and complemented nationalist concerns: the preservation of Muslim identity and rejuvenation of the Islamic community in the face of the threat of political and cultural assimilation; the achievement of Muslim unity and solidarity to attain autonomy and independence. In North Africa, Iran, and South Asia, Islam provided a common identity and allegiance, ideology and symbols, leadership and mosque-based centers for organization and communications. Traditional calls to the defense of Islam (Islam in danger!) and "Allahu Akbar" (God is most great) proved effective rallying cries among many. Islamic leaders and organizations were often key actors in independence and nationalist movements.

The Arab East

Islam played a less prominent role in the early development of Arab and Egyptian nationalism. The first stirrings of Arab nationalism emerged in reaction to Ottoman rule rather than as a response to the West. The Arab Christian Literary Movement's emphasis on Arabic language, with its sense of a community identity rooted in literature and history rather than in religion, and the Young Turks' Turkification program, which shifted emphasis from Ottoman to Turkish ethnic and linguistic identity, fostered nascent nationalist sentiments among Arab and religious and ethnic minorities (Balkan, Greek, Syrian) over issues of language, identity, and political autonomy during the late nineteenth and early twentieth centuries.[24] Traditional transnational Islamic and imperial or sultanate forms of government experienced the stirrings of modern forms of nationalism based on common language, territory, and ethnic ties.

However, Arab and Egyptian nationalism did not really take shape and function as an effective tool in the struggle for independence from European hegemony until after World War I. Two factors were particularly important, reflecting both the persistent threat of the West and the growth of Muslim resistance. In 1920, in retribution for Ottoman support for Germany during World War I, the Allies broke up the Ottoman empire and carved out a number of modern nation-states. Under the Treaty of Sèvres Britain and France set up a mandate system which gave Britain control over Palestine (including modern-day Jordan) and Iraq, while France governed Syria (including modern-day Lebanon), and the Hejaz (part of Saudi Arabia) remained independent. At the same time, the struggle against European imperialism intensified with the development of nationalist movements of independence within which Islam was a factor to varying degrees.

While Arab and local nationalists often found it necessary to acknowledge their Islamic heritage, Islam was subordinated to more secular forms of nationalism whose primary emphasis was on other factors such as common language, history, traditions, and territory. Thus for example, in Egypt former students of Afghani and Abduh like the Egyptian nationalist leader Saad Zaghlul (d. 1927) and Taha Husayn pursued a more secular Egyptian nationalist path.

This secular turn of Afghani and Abduh's disciples was countered by Rashid Rida (d. 1935), who carried on the reformist tradition of Muhammad Abduh. However, the political realities of the post–World War I period pushed Rida toward a more conservative and anti-Western position. While maintaining the transnational Islamic ideal and identity of Muslims, he pragmatically and reluctantly accepted the reality of modern states and nationalism, though he would subordinate them to membership and solidarity in the broader Islamic community. The progressive secular bent of Egyptian modernists, especially former students of Afghani and Abduh, influenced Rida's growing admiration for Saudi Arabia's self-proclaimed Islamic state and his alignment with religious leaders (*ulama*) whose conservatism he had previously criticized. Thus Rida in later life reacted to the secularization and Westernization of Muslim society, which he saw as outcomes of modern rationalism. He increasingly moved away from Islamic modernism's openness to cultural synthesis. Instead, he emphasized the self-sufficiency and comprehensiveness of Islam and took a more critical attitude toward the threat of the West. Rida's more defensive and polemical posture both reflected the failure of Islamic modernism to secure a firm middle ground between rejection and cultural assimilation, and heralded the appearance of activist Islamic organizations—the Muslim Brotherhood in Egypt and the Jamaat-i-Islami (Islamic Society) in Pakistan—which possessed a more anti-Western and consciously self-sufficient Islamic ideology.

The Arab West (North Africa, Maghreb)

In North Africa Islamic reformism, inspired ideologically by Afghani and Abduh's Salafiyya movement, became closely allied with the nationalist movement after World War I. North African reformers countered French rule and the dangers of cultural assimilation, preached the need for a revival of their Arabic and Islamic identity and heritage, and advocated a modernist interpretation of Islam. Islamic reformers were among the founders and leaders of early nationalist organizations and parties which emphasized independence and a national identity based upon an Arab–Islamic heritage: Allal al-Fasi who led Morocco's religiously oriented Istiqlal (Independence) party in 1946; Abd al-Aziz al-Thalibi, founder of Tunisia's Destour (Constitution) party after World War I; and Abd al-Hamid Ibn Badis (Ben Badis, 1889–1940), who organized the Association of Algerian Ulama in 1931.

Religion was employed by religious and secular-minded Muslims alike who recognized its ability to transcend local religious (sectarian and Sufi) differences, and ethnic and tribal differences, to achieve mass mobilization. Islam was used as a symbol of national unity and identity with respect to French rule and the threat of cultural assimilation. It proved an effective means to mobilize popular support and to bridge the gap between Arab and Berber in Morocco and Algeria.

Iran

Iran differed from most of the Arab world and South Asia in retaining its independence. However, it remained vulnerable to British and Russian imperial designs: the British in southern Iran and the Russians in the north threatened Iranian sovereignty. Two events proved formative in the development of Iranian nationalism, the Tobacco Protest (1891–92) and the Constitutional Revolution (1905–11). Each represented a different response to Western influence. The former was motivated by the desire to protect Iran from foreign infiltration and economic dependence, and the latter by the need to limit the arbitrary powers of the Shah through Western-inspired constitutional reform.

When the Shah in 1890 gave a British company a monopoly over the sale and export of tobacco, a nationwide boycott and strikes were organized, led in large part by merchants and Muslim religious leaders. The protesters received support from Jamal al-Din al-Afghani, an adviser to the Shah, who warned that economic concessions were a major step toward foreign rule. Ayatollah Hasan al-Shirazi, a leading cleric, issued an Islamic decree or ruling (*fatwa*) which forbade smoking, and leading mosques provided sanctuary for protesters. Fearing Russian intervention, the Shah gave in to the protesters' demands.

The success of the Tobacco Protest and the desire to limit the arbitrary acts of their rulers emboldened religious leaders, using their mosques as centers of resistance and asylum, to again join with other sectors of society—landlords, merchants, army officers—in the Constitutional Revolution. Although the Shah initially resisted demands for a Western-style constitution, public protests as well as pressure from Britain and Russia led to the creation of a national assembly in 1906. However, Iranian nationalist gains evaporated over the following years when a change of rulers occurred, and Britain and Russia were able to carve out their own spheres of influence. Yet as a result of its experience in the late nineteenth and early twentieth century, Iranian nationalism had begun to counter foreign influence and domination with a strong reaction that included an Islamic component.

South Asia

In southern Asia both Pan-Islamic and Muslim nationalist movements developed in reaction to European colonialism. Despite sporadic anti-British Muslim riots, Muslims did not follow the Ottoman empire's alignment with Germany during World War I but instead generally remained pro-British. However, in the aftermath of the war, concern in the Muslim world that the Allies would dismember the Ottoman empire and abolish the caliphate led to a growth of Pan-Islamic sentiment which in the Indian subcontinent expressed itself in the formation of the Caliphate Movement in 1919. Within a year of its founding, Muslim fears were realized with the partition of the Ottoman empire in 1920, which ushered in the mandate period. The Caliphate Movement continued until 1924, when Kemal Ataturk, the ruler of Turkey, formally abolished the caliphate. However, its spirit of anti-European imperialism and national independence was carried over into the Indian National Congress, the major nationalist movement in India, whose membership cut across communal boundaries and thus included leaders from all of India's major religious groups.

By the thirties, however, Muslim separatism had increased. Growing conflict between Hindu and Muslim communities and the results of provincial elections in 1937 convinced Muhammad Ali Jinnah and Muhammad Iqbal of the Muslim League that the tendency in India to vote along communal lines threatened the identity of the Muslim community and might well lead to its domination by the Hindu majority in an independent India. The Muslim League increasingly demanded a separate Muslim state and became the chief proponent of Muslim nationalism. Despite Jinnah's secular disposition and Western life-style, the more secular-oriented leadership of the Muslim League placed greater emphasis on Islam so as to overcome its lackluster track record. Islamic slogans and symbols proved effective in mobilizing disparate Muslim groups in the demand for a Muslim homeland.

Ironically, the Islamic politics of the Muslim League proved surprisingly ineffective among religious leaders. Most religious leaders refused to support the League's call for a separate Muslim state, maintaining that nationalism and Islam were antithetical. The reasons for their condemnation of nationalism varied, influenced by anti-Europeanism and local politics as much as religious belief. Nationalism was regarded as a Western concept whose narrow particularism was contrary to Islamic universalism and divisive. Abul Hasan Ali Nadwi, a religious scholar and community leader, maintained that nationalism was "a narrow national feeling, racial prejudice, and an exaggerated regard for geographical division [which] are the characteristics of the Western mind."[25] Mawlana Abul Ala Mawdudi, another prominent religious leader and the founder of the Jamaat-i-Islami, noted: "Be it in the sphere of economics or politics, or civics or legal rights and duties, those who accept the principles of Islam are not divided by any distinction of nationality, of class, or country."[26]

Although much has been made of Islam's incompatibility with nationalism, the objections of Indian religious leaders to Muslim nationalism were motivated as much by their opposition to the leadership of Muhammad Ali Jinnah, whom they considered part of the Western secular elite, as by ideological conviction. Most were convinced that the Muslim nationalism of the Muslim League was motivated solely by a hypocritical attempt to win votes, and that it disguised Jinnah's intention to create a secular rather than Islamic state in the newly proposed Muslim homeland of Pakistan.

The peoples of the Indian subcontinent finally gained independence in 1947. Two states were carved out: a secular India that was overwhelmingly Hindu but with a significant Muslim minority, and Pakistan, a Muslim homeland where Muslim nationalism, it was hoped, would unite West and East Pakistan (presently Bangladesh), which were separated by a thousand miles of Indian territory.

Emerging States and the West: From Emulation to the Failure of Liberal Regimes

In the aftermath of independence, the relationship of Islam to the West was less one of conflict than of grand emulation. As newly emerging states struggled to establish themselves, the West proved a necessary and often popular source and model. Although the independence struggle left deep resentment and scars, most rulers appropriated their colonial institutional legacy and ties. Modernization was imposed from above by governments and Westernized elites. European languages remained the second (and, among modern elites, often their preferred) language. In some countries European languages were the official language of government, the courts, and university education. Modern bureaucratic, educational, and legal systems continued intact, as did trade and commerce. Islamic law was generally confined to the area of personal status or family law. Thus, for example, French language and culture remained influential in North Africa and Lebanon, and English in southern Asia and Malaysia. Individuals, countries, cities, and institutions judged themselves, and were judged, to be modern by the degree to which they were Westernized—in language, dress, manners, knowledge, organizational structure and values, architecture, and infrastructure. America, in particular, enjoyed a certain pride of place, since it lacked the negative baggage of European colonial powers.

The Western orientation of development and modernization theory and its application in the Muslim world set the stage for decades of educational and technical exchange as well as political, economic, and military alliances. Few questioned the accepted wisdom that modernization meant the progressive Westernization and secularization of society. Admiring the West for its democratic ideals, military power, and technology, students in many Muslim countries prized a modern Western education at home and especially abroad. Such an education was often the surest ticket to responsible positions in government, business, and academia. Close international ties were formed between governments, the military, oil companies, and banks. Thus, for example, Habib Bourguiba, the "Great Combatant" exiled by France during the nationalist struggle, increasingly aligned Tunisia with France and French culture. French, not Arabic, was the language of higher education and culture. King Hussein of Jordan and the Shah of Iran became close allies of the United States. Pakistan remained a member of the British Commonwealth, and English served as the language of government and higher education.

However, during the fifties and sixties widespread dissatisfaction with the track record of Western-inspired liberal nationalism took its toll, as monarchs and governments tumbled from power in Egypt, Libya, the Sudan, Iraq, and Algeria. Egypt's growing disillusionment with Europe was not atypical.

> World War I, the political strife of the inter-war years, the Great Depression, and finally, World War II exposed Egyptians to the failings of the constitutional and liberal governments of Europe, the ruthlessness of the powers, their indifference to principle in manipulating non-European peoples, and their contempt for their subjects. These events shook the confidence of many Egyptians and the future belonged to the West, and left others disappointed and alienated. Furthermore, the failure of the liberal regime . . . to deal equitably with the country's political and economic

problems also undermined faith in parliamentary regimes and in the value of individualism.[27]

Two ideological orientations or movements emerged, both of them populist: the Islamic activism of the Muslim Brotherhood and the Arab nationalism/socialism of Gamal Abdel Nasser. Both emphasized indigenous Arab–Islamic roots and sources, stressed Arab unity, and were critical of the failures of liberal nationalism and the West. Both the Brotherhood and Nasser would not only capture the imagination of Egyptians but also impact the Arab world and beyond. While Arab nationalism attempted to subsume Islam, Islamic activism asserted the primacy of Islam and called for an Islamic order—a political system guided by Islamic law—as the basis for Arab unity and solidarity.

Modern Islamic Neorevivalist Movements and the "Failure of the West"

Islamic reform and Muslim responses to European colonialism and the West took a significant ideological turn in the thirties and forties with the creation of two modern Islamic organizations, the Muslim Brotherhood of Egypt and the Jamaat-i-Islami (Islamic Society) in the Indian subcontinent. Both embodied a growing ambivalence toward the penetration of Western culture, the threat of secular nationalism, and the continued political presence of Western imperialism.

> The West surely seeks to humiliate us, to occupy our lands and begin destroying Islam by annulling its laws and abolishing its traditions. In doing this, the West acts under the guidance of the Church. The power of the Church is operative in orienting the internal and foreign policies of the Western bloc, led by England and America.[28]

Both the Brotherhood and the Jamaat emphasized Islam's ideological self-sufficiency, and were less accommodationist, and far more critical of the West. Whereas Islamic modernism had sought to learn from and emulate the success of the West, the Muslim Brotherhood and the Jamaat emphasized the failure of both the West (capitalism) and the East (Marxism) as models for development in the Muslim world. They denounced the Westernization and secularization of Muslim societies, the divisiveness of nationalism, and the excesses of capitalism, as well as the materialism and godlessness of Marxism. Muslims were told to remember that they possessed a third way, an alternative to foreign models and systems—Islam.

Their desire to transform society invariably led to involvement in politics and, at times, confrontation with their national governments. Activists and national regimes accused each other of violence and sedition. At various times the activist leaders were arrested and their organizations suppressed. Hassan al-Banna was assassinated in 1949, and Brotherhood leaders were executed and the Muslim Brotherhood officially repressed and dissolved in the late sixties. While Mawlana Mawdudi and Jamaat leaders were imprisoned and even condemned to death on one occasion, the Jamaat-i-Islami was able to participate in the political process more freely than their Egyptian counterparts.

The significance of the Muslim Brotherhood and the Jamaat-i-Islami extended far beyond their national homelands and in time took on transnational significance. The Brotherhood inspired the establishment of similar organizations in the Sudan, Syria, Jordan, the Gulf, and Africa. The Jamaat developed sister organizations in India, Bangladesh, Afghanistan, and Kashmir. The writings of the Brotherhood's Hassan al-Banna and of Sayyid Qutb and Mawlana Mawdudi of the Jamaat-i-Islami would in time become widely translated and disseminated throughout much of the Islamic world. Their vision of Islam as an alternative ideology for state and society and the example of their organizations and activities provided a model for future generations of Muslims. As such, for many they constituted a link between the traditional religious heritage and the realities of modern life.

Arab Nationalism/Socialism and the West

Arab nationalism/socialism and the figure of Gamal Abdel Nasser dominated much of the Arab world and challenged Western powers from the midfifties on. As the influence of European colonialism diminished, the United States and the Soviet Union emerged as superpowers. Western capitalism, Marxism, and socialism were contending forces in a Cold War that seemed to force developing nations to choose between the West or communism. Liberal nationalism seemed exhausted or discredited as regimes were toppled by Arab socialist movements. New governments advocating Arab socialism came to power in Egypt, Syria, Iraq, Libya, the Sudan, and Algeria. Nasserism, Egypt's Gamal Abdel Nasser's Pan-Arab nationalism/socialism, inspired the revolutions of Libya's Muammar Qaddafi and the Sudan's Gaafar Muhammad Nimeiri in the late sixties; the Arab socialism of the Baath ("resurrection") party in Syria and Iraq; and the Arab–Islamic socialism of Algerian independence (1962).

All new governments shared a worldview prevalent among many Third World countries. They indicted European imperialism for its invasion and occupation, as well as its policies which divided by creating states and drawing artificial national boundaries, thus debilitating the Arab and Muslim world. They espoused a continuing struggle against colonialism, exacerbated by the West's role in the creation and support of a Western colony, Israel, in the Arab homeland, and denounced the failures of the traditional Arab political leadership and its Western liberal nationalism. They condemned the radical individualism of capitalism, called for Arab unity and solidarity, and promised the creation of a new social order to alleviate the plight of the masses in Arab societies.

Perhaps no modern Arab leader captured the imagination of the Arab world and the Third World more than Gamal Abdel Nasser (1918–70). Nasser's memory cast a long shadow over political development in the Middle East. His name came to be associated with the two confrontations that epitomized the anti-imperialist mood and politics of the period: Suez and Palestine. A charismatic leader whose ability to sway a crowd was the envy of every Arab politician and the bane of his enemies, Nasser fired the imaginations and mobilized the emotions of Arabs and Muslims far beyond the borders of Egypt. Wishing to establish his leadership throughout the Arab world, Nasser broadened Egyptian nationalism into an Arab nationalism/socialism rooted in the region's common Arab–Islamic heritage. Although he competed with the Hashemites of Iraq and Jordan as well as the House of Saud for leadership in the Arab world, among the peoples of that world he was unrivaled in popularity, the object of a personality cult matched by very few Arab or Muslim rulers except perhaps the Ayatollah Khomeini. There were Nasserite parties in Arab countries and active student organizations in Lebanon, Syria, and Jordan. Nasser's picture adorned the walls of homes and the streets of cities and villages in many countries. He was seen as the savior of the Arabs and Palestinians.

Nasser, the man and his message, had many facets. He will long be remembered for his call for Arab unity, his defiance of the West and assertion of Arab independence from Western dominance in the Suez. He asserted Egypt's claim to leadership of the Arab world, championed the struggle for the liberation of Palestine, and espoused "positive neutrality" in the Cold War between the United States and the Soviet Union. He became the architect of Arab socialist reform.

The overthrow of the British-dominated regime of Egypt's King Farouk by Nasser and the Free Officers in the July 1952 coup was cast as a repudiation of European colonial influence, the failed policies of Western liberal nationalism and capitalism, and the ignominy of Egypt's defeat in Palestine in 1948. Nasser's nationalist and anti-imperialist (often interpreted as anti-Western) credentials and his transnational popularity and leadership in the Arab and Third worlds were solidified by his defiance of the West over control of the Suez Canal and by his creation, with India's Nehru and Indonesia's Sukarno, of the Alliance of Nonaligned Nations.

When the United States withdrew its promised loan for building the Aswan Dam in 1956, Nasser seized control of the Suez. He nationalized the Suez Canal Company, dramatically proclaiming to a tumultuous crowd: "O Americans, may you choke to death in your fury!"[29] He stood his ground as Britain's Prime Minister Anthony Eden denounced him as a Hitler (much as George Bush would later denounce a would-be successor to Nasser, Saddam Hussein). Assisted by Israel, which the Arabs regarded as a colony of Europe and the United States, France and Britain invaded Egypt in 1956, while Israel occupied the Sinai. Although it was U.N. intervention and pressure from the United States against this violation of international law that forced Britain, France, and Israel to withdraw, Nasser was credited in the Arab, Muslim, and broader Third World with this singular defeat of European colonialism. In the eyes of many Western governments, Nasser's defiance of the West at Suez—coupled with his impassioned denunciation of Western imperialism, his wars with Israel, and his doctrine of "positive neutrality" (he accepted arms from the Soviet Union and recognized the People's Republic of China)— branded him as simply anti-Western. However, for much of the Third World he became the symbol of anti-imperialism.

The emergence of the Arab nationalism/socialism of Nasser and the Baath party signaled a period in which local or state nationalism was transformed into or equated with a transnational, Pan-Arab nationalist sentiment that stressed Arab political unification and independence from foreign domination. Both Nasserism and the Baath blamed European colonialism and the rivalry for power among pro-Western Arab monarchs for the ills of the Arab world. The artificial façade of modern nation-states and societies, whose boundaries were often arbitrarily drawn by Europe and whose European-appointed or dependent leaders lacked political legitimacy, had produced a fractured Arab nation. These divisions, they argued, had perpetuated Arab weakness and dependence upon the West and rendered the Arab world incapable of thwarting the creation of Israel.

The program of Arab nationalism was idealistic, revolutionary, and ambitious. It affirmed an ideal rather than a reality—the existence of an Arab nation that possessed a political and economic unity. Arab unity and solidarity were rooted in a transnational Arab identity based upon common language, history, and territory. Arab socialism promised the creation of a new social order, based upon state planning and control of major industries, financial institutions, and utilities. Arab socialism was to be a third alternative to capitalism and communism, eschewing the capitalist evils of individualism (the concentration of economic power and wealth) and consumerism as well as Marxism's atheism and theory of class conflict. Arab regimes were denounced for their failure to respond to the needs of the predominantly poor masses of workers and peasants. This failure was attributed to economic systems controlled either by the indigenous feudalism of a few privileged elites or by capitalism, which was equated with Western ownership or economic dependence. Arab socialism promised a more socially just society through state control of national resources and production, a more equitable distribution of wealth, and social services.

The union of Egypt and Syria in 1958 through the formation of the United Arab Republic, with Nasser elected its first president, and a Baathist coup in Iraq in 1963 seemed to signal the incorporation of these three major countries in a new Arab order. It was, however, a short-lived promise, for Nasser and the Baathists failed to resolve differences over leadership of the new state. But the failure of Arab leaders and of Arab socialism to create a transnational political union was offset by their rhetorical and political opposition to Israel. Palestine proved an effective symbol and rallying post.

Palestine

The creation of Israel in 1948 was seen as the boldest example of the duplicity of European colonialism and its desire to keep the Arabs divided and weak. Israel itself was considered a European–American colony in the midst of the Arab nation. The Arab defeats in the 1948 and 1956 wars were a further humiliation. For Arab leaders, Palestine provided a no-lose cause (not threatening class, political, or religious interests) that each could exploit domestically and internationally as rulers competed in the force of their denunciations. Military leaders and monarchs, the educated and the uneducated, the landed and the peasants, the secular and the Islamically oriented—all could identify with the plight of the Palestinians. The struggle against Israel symbolized the battle against imperialism, provided a common cause and sense of unity, and distracted from the failures of regimes and of Arab nationalism/socialism. As in the nationalist struggle, both the secular and the religiously oriented, Arab nationalists and Islamic activists, found common ground in regarding the liberation of Palestine as a great jihad against Western imperialism.

Arab Nationalism and Islam

While Arab nationalism was primarily secular in orientation, it did acknowledge the Islamic component of Arab identity and history. Arab nationalism had an affinity (linguistic, historical, and religious) with Islam. Arabic is the language of revelation and official prayer in Islam; Arab and Islamic history are intertwined (early heroes, conquests, and accomplishments); the homeland and early centers of power and culture were located in the Arab world. Moreover, Arab nationalists and the Muslim Brotherhood, active in Egypt and Syria, shared common concerns such as anti-imperialism, the need for Arab unity and solidarity, and the liberation of Palestine. However, although the Muslim Brotherhood in Egypt had initially backed Nasser and the Free Officers, the Brotherhood went over to the opposition when it became clear that Nasser did not intend to establish an Islamic government. Nasser's government clashed violently with the Brotherhood and suppressed it in 1954 and again in 1965, after abortive attempts to assassinate Nasser; thousands were imprisoned, and a number of its key leaders executed.

   Though Nasser would brook no Islamic opposition, the realities of Egyptian and Arab society caused him to increasingly use or manipulate religion to legitimate his state socialism and broaden his popular support. His need to counter and control an Islamic opposition as well as to resist Saudi Arabia's increasing assertion of transnational Islamic leadership moved Nasser to use Islam carefully in his foreign policy. The Saudis had become progressively more leery of Nasser's aggressive, expansionist assertion of Pan-Arab leadership. They were also threatened by Arab socialism's populist critique of the "feudalism" of conservative Arab monarchies and its support for their overthrow. The Saudis countered the threat of Arab nationalism by espousing a Pan-Islamic ideology and leadership and by condemning Nasser's un-Islamic "socialism." They used their Islamic claims as keepers of Islam's sacred cities of Mecca and Medina and protectors of the pilgrimage, and their oil wealth, to promote themselves as patrons of Islam and to encourage Muslim solidarity. They created international Islamic organizations such as the World Muslim League (1966) and the Organization of the Islamic Conference (1969), through which they distributed funds for the promotion and preservation of Islam (the building of mosques, schools, hospitals, and the printing and distribution of religious literature) and organized Muslim countries and institutions.

Nasser countered by employing religious symbols, leaders, and institutions to legitimate and win support for his Arab socialist ideology and policies. His government controlled or coopted the religious establishment. It nationalized al-Azhar University, the oldest seat of Islamic learning and religious authority, and controlled salaries and funding for many mosques and religious personnel. Nasser used both the religious scholars of al-Azhar and a state magazine, *The Pulpit of Islam,* to legitimate Arab socialism and its policies of state ownership and nationalization. He identified Islam with Arab socialism and its principles of anti-imperialism, social justice, and equality. At a meeting of the Arab–Islamic conference on liberation organizations, Nasser emphasized that the Arab and Islamic worlds were facing one enemy—imperialism. Egypt sponsored international meetings such as the Afro-Asian Conferences in 1964 and 1965, which explored Islam's relevance to the struggle against imperialism.[30]

However, at its core Nasser's Arab nationalism, like that of the Baath, remained primarily secular in orientation. Nowhere was this clearer than in the 1967 Arab–Israeli war, which was fought under the banner and slogans of Arab nationalism/socialism. The disastrous defeat of Arab forces stood as an indictment of Arab nationalism, further inflamed Arab and Muslim passions against Israel and American neoimperialism, and became a major catalyst for the Islamic resurgence.

The first half of the twentieth century witnessed both the end of European colonialism and the emergence of modern Muslim nation-states. Both experiences were heavily dependent upon or influenced by the West. The breakup of the Ottoman empire and the creation by Britain and France of mandate countries created a legacy which has contributed to the instability of regimes and leaders down to the present. It enhanced the image of a militant imperialist West and created modern states whose artificially drawn boundaries and mandate-appointed or approved rulers possessed questionable political legitimacy.

Throughout the turbulent politics of the postcolonial, postindependence period, the roller-coaster relationship between the West and the Muslim world continued. On the one hand, Europe and America represented the example par excellence of modernization and development, of what it meant to live fully in the modern world, from architecture and medicine to education and technology. On the other hand, in the eyes of many, the Muslim world's relative backwardness and Sisyphean battle to build strong, stable societies were attributed to the bitter harvest of European colonialism. The creation of Israel, the Suez crisis, and the politics of the Cold War were regarded as aspects of post–World War II neocolonialism, a hegemonic chess game between the United States and the Soviet Union which threatened the identity and integrity of the Muslim world.

While Arab nationalism/socialism had swept aside the failures of liberal nationalism, the heady visions of unity and power symbolized by Gamal Abdel Nasser in the sixties were shattered by the inescapable realities of the 1967 Arab–Israeli war. Arab nationalism/socialism had proved a grand bust, a shattered myth—in the eyes of others, a setback from which it still has not recovered. Ideologically, it failed to produce Pan-Arab unity and solidarity: Pan-Arabism proved incapable of transcending the diverse and often competing interests of Arab leaders and societies. Socioeconomically, Arab socialism with its nationalization and land reform did not usher in the promised egalitarian Arab order or significantly alter the plight of the masses. As the Six-Day War demonstrated, diplomatically and militarily the Arab governments had proved impotent in the international arena. Parliamentary democracy had proven incapable of controlling ruling elites and the military. Despite several decades of independence, imported ideologies and development schemes seemed to have failed. Old issues of identity, national ideology, political legitimacy, socioeconomic reform, and Western domination persisted. The failures of governments and national ideologies (liberal nationalism and Arab socialism), epitomized by the humiliation of 1967, precipitated a deepening sense of disillusionment and crisis in many Muslim societies and contributed to the political and social resurgence of Islam.

4

Islam and the State:
Dynamics of the Resurgence

In the minds of most Westerners, for whom the history of Islam and the West are at best a vague notion, knowledge of and attitudes toward Islam and the Muslim world have been shaped by contemporary images and experiences encapsulated in terms such as *Islamic fundamentalism* and *terrorism*. The phrase "Muslim fundamentalist" has become a convenient, if misleading, way for the media and Western governments to identify a wide-ranging array of groups in the Muslim world, as well as a ready "bogey" for Muslim regimes wishing to discredit and disparage their opposition. For many, "fundamentalism" conjures up images of mobs shouting death to America, embassies in flames, assassins and hijackers threatening innocent lives, hands lopped off, and women oppressed. The contemporary revival of Islam in Muslim politics is far more multifaceted and significant than these images and slogans communicate. Its presence in and impact upon Muslim societies is more pervasive and nuanced as both a political and a social phenomenon. What is the diversity of political Islam as revealed in the recent experiences of Muslim states? How have governments used Islam, and how has the implementation of Islam affected the dynamics and development of Muslim politics? How has the West misinterpreted, and developed policies that have reinforced, Muslim stereotypes of an anti-Islamic neocolonialism and thus contributed to further alienation and resentment?

Islam and the Modern State

By the midtwentieth century most of the Muslim world had achieved political independence. The influence and continued attraction of the

77

West was evident in the more secular path chosen by most rulers and modern elites. Even in those countries where Islam had played an important role in nationalist movements, the new generation that came to power opted for a more secular orientation. As one looked across the Muslim world, three orientations or models for the relationship of religion to the state could be distinguished: Islamic, secular, and Muslim. Saudi Arabia was a self-declared Islamic state. The monarchy of the house of Saud based its legitimacy on Islam, claiming to govern and be governed by the Quran and Islamic law. The house of Saud had established a close relationship with the *ulama,* who continued to enjoy a privileged position as advisers to the government and officials in the legal and educational systems. The Saudi government used Islam both to legitimate domestic policies and to conduct foreign policy.

At the other end of the spectrum Turkey, the only remnant of the Ottoman empire, had opted for a secular state with religion severely restricted to personal life. Turkey, under the leadership of Kemal Ataturk (president, 1923–38), embarked upon a comprehensive process of Turkification and Westernization, and a secularization that transformed language and history as well as religion and politics. Western script replaced Arabic script, and history was rewritten, suppressing its Arabic component and glorifying (and at times fabricating) its Turkic heritage. Ataturk autocratically oversaw a series of reforms which deposed the sultan, abolished the caliphate, disestablished Islam, closed seminaries, proscribed the wearing of clerical garb, and replaced traditional institutions (law, education, government) with modern, Western-inspired alternatives.

The majority of countries in the Muslim world, however, fall into a middle position. They are Muslim states, in that the majority of the population and its heritage are Muslim, yet they have pursued a moderated secular path of development. While most looked to the West for the basis of their systems of modern constitutional government, law, and education, they have also injected Islamic provisions into their constitutions, requiring that the head of state be a Muslim or that Islamic law be recognized as a source of law (even when this was not the case in reality). These governments sought to control religion by incorporating religious institutions within the bureaucracy—within ministries of law, education, and religious affairs. With few exceptions the general trend, expectation, and goal of governments and their Western-educated elites was to create modern states modeled upon Western paradigms.

Ideologically the prevailing tendency, following Western models, was to foster secular forms of national identity and solidarity and to limit religion to private rather than public life. Local (Egyptian, Syrian, Libyan) or regional/linguistic (Arab or Baath) forms of nationalism or socialism prevailed. This secular trend began to change almost imperceptibly in the sixties, however, and the change became more pronounced in the seventies and eighties. As examples from the Sudan, Egypt, Libya, and Iran illustrate, the usage and manifestations of Islam varied significantly. Differences in sociopolitical environments, leadership, and economic conditions determined how Islam was defined and implemented. Similarly, Islam proved to be both a challenge and a threat—a source of stability and instability, of legitimacy and revolt, utilized by pro-Western and anti-Western governments alike. Military (and ex-military) rulers like Libya's Muammar Qaddafi, the Sudan's Gaafar Muhammad Nimeiri, Egypt's Anwar Sadat, and Pakistan's Zulfikar Ali Bhutto and Gen. Zia ul-Haq appealed to Islam to strengthen their legitimacy, mobilize popular support, and justify government policies. By contrast, Tunisia's Habib Bourguiba and the Shah of Iran pursued more overtly secular paths.

Sadat, Qaddafi, and Nimeiri shared a common link with Egypt's charismatic leader, Gamal Abdel Nasser. Qaddafi and Nimeiri seized power in 1969, patterned their revolutions on Nasser's 1952 July Revolution, and aligned themselves with Nasser's brand of Arab nationalism/socialism with its vision of Arab unity. In 1970 Qaddafi, Nasser, and Nimeiri concluded the Tripoli Pact, which aimed at the complete merger of Libya, Egypt, and the Sudan. A similar pact was concluded with Nasser's successor, Anwar Sadat, but it ultimately failed to be implemented.

In the wake of the 1967 Arab defeat and its discrediting of Arab nationalism, the revolutionary nationalist/socialist governments of Libya, the Sudan, and Egypt turned to Islam to buttress their faltering nationalist ideologies.

Libya

> O you people, tear to shreds all imported books which do not set forth [the values of] the Arab heritage and of Islam, of socialism and of progress.
>
> —MUAMMAR QADDAFI

With these striking words, Muammar Qaddafi announced his wide-ranging cultural revolution, an indigenous socialist vision rooted in Libya's Arab heritage and Islamic faith.[1] Few leaders in the Muslim world have captured the attention of the West and symbolized "Islamic radicalism and terrorism" in the seventies and eighties more stridently than Qaddafi. Long before the Islamic revolutionary slogans of Khomeini's Iran, billboards in Libya, referring to Qaddafi's seizure of power on the first (Al Fateh) of September, declared: "Al Fateh is an Islamic Revolution."[2]

Libya's modern development has been profoundly affected by two events. In 1959 oil was discovered there, and so this poor nation of one million inhabitants, which had been economically dependent upon Britain and America, became a leading oil producer within a decade. In 1969 a dramatic coup d'état brought Qaddafi to power. As with similar revolutionary movements in the Arab world that came to power in Egypt, Syria, Iraq, and Algeria during the late fifties and early sixties, the rationale for the Libyan coup was socioeconomic reform necessitated by the failure of a Western-influenced monarchy. Under Qaddafi three interrelated ideas were woven into Libya's ideological identity: Pan-Arabism, socialism, and Islam.

Qaddafi's appeals to Islam were influenced by his personal piety and the sociopolitical realities of his country. "Islam has been and remains the basic glue of Libyan society. As the primary unit of loyalty and identity, religion has been 'a political symbol of crucial importance in controlling and mobilizing the masses.' "[3] Shortly after coming to power in September 1969, Qaddafi turned to Islam so as to enhance his legitimacy and that of Libya's socialist revolution as well as to spread his influence in the Arab and Muslim world.[4]

Qaddafi deposed the conservative monarchy of King Idris, whose right to rule had been Islamically legitimated by his descent from the nineteenth-century Islamic revivalist leader Muhammad ibn al-Sanusi (1787–1859), who had established an Islamic state in Libya. Thus it was not surprising that Qaddafi and the clique of young officers who had overthrown the king found it useful, if not necessary, to justify their actions Islamically. Qaddafi's early statements placed his new government on an Islamically legitimated Arab socialist path: "a socialism emanating from the true religion of Islam and the Noble Book."[5]

During the seventies Qaddafi introduced a series of reforms which reasserted Libya's Arab–Islamic heritage and thus employed Islam to buttress its national ideology and Arab nationalism/socialism. This emphasis combined his fundamental belief in both Arab nationalism and Islam, in both Arab and Muslim unity. Signs of Libya's European (Italian, British, and American) Christian colonial past were suppressed or replaced. Churches were closed; missionary activities were banned; British and American military bases were closed. Arabic was reintroduced as the official language; Arabic names and street signs replaced their Western counterparts. Islamic laws forbade alcohol consumption, nightclubs, and gambling; the Islamic Call Society was established for the propagation of Islam at home and abroad. Islamic criminal penalties (the *hudud,* "limits" of God) were reinstated: amputation for theft, stoning for fornication and adultery. However, if closely examined, despite much fanfare by the Libyan government and the Western press, Islamic law has in fact remained relatively peripheral to Libyan society.

During the latter part of the seventies, Qaddafi issued his master plan, a utopian vision of society, *The Green Book.* Its title was deliberately suggestive in its Islamic and universal claims. Muslims believe that their book, the Quran, is the literal word of God and source of guidance for Muslim society. They acknowledge that Jews and Christians as People of the Book (i.e., recipients of God's revelation). The Chinese had their *Red Book,* Mao Tse-tung's ideological guide for Third World revolution. Since the color green is associated with the Prophet Muhammad, the title of *The Green Book* could be understood as both an Islamic alternative and an option for the Third World as well. Similarly, just as China had its Red Guard, Qaddafi also created a Green Revolutionary Guard to implement his cultural revolution.

The Green Book is a series of three small volumes: *The Solution to the Problem of Democracy* (1975), *Solution of the Economic Problem: Socialism* (1977), and *Social Basis of the Third International Theory* (1979). It was taught in the schools, became required reading for all Libyan citizens, and was propagated through the innumerable wall posters which seemed to dominate the Libyan landscape. Qaddafi's assessment of the Arab world's plight was cast in the categories of traditional Islamic revivalism. *The Green Book* championed many of the themes common to Arab nationalism and contemporary Islamic thought: anticolonialism and anti-imperialism; dependence on the West as the source of Muslim weakness and loss of identity; social injustice characterized by exploitation and corruption; the return to Islam to restore Arab/Muslim power and greatness; and Qaddafi's own brand of Islamic socialism or social justice. Qaddafi called for return to an Islam that provided a comprehensive guide for humanity: "We must take the Quran as the focal point in our journey in life because the Quran is perfect; it is light and in it are solutions to the problems of man . . . from personal status to . . . international problems."[6]

The Green Book heralded a universal, revolutionary ideology, Qaddafi's Third International Theory, for the creation of a new political and social order—a third alternative to the two great modern options of capitalism and Marxism. Despite the initial Islamic ideological basis for *The Green Book,* the last two installments omitted any reference to Islam. Qaddafi's ambitions were far broader. First, *The Green Book* was to serve as the blueprint for Libyan society, which would then serve as a model for the Third World. Second, his revolutionary goal was to create a new populist socialist world order with *The Green Book* as its guide: "*The Green Book* is the guide to the emancipation of man. *The Green Book* is the gospel. The new gospel. The gospel of the new era, the era of the masses."[7]

Qaddafi's new gospel for the world was but an extension of his earlier decision to implement his own version of a socialism distinct from the prevailing foreign ideologies and erected on the twin pillars of Arabism and Islam:

> Our socialism is both Arab and Islamic. We stand midway between socialism and communism and socialism and capitalism. Our socialism springs directly from the needs and requirements of the Arab world, its heritage, and the needs of society. It consists of a social justice which means sufficiency in production and just distribution. These principles are to be found in the Islamic religion, and particularly in the law of *zakat* [alms].[8]

Qaddafi implemented his vision for society during the midseventies, creating a populist socialist state of the masses (*jamahiriyah*). Libya's new populist identity and ideology were institutionalized in March 1977 when the General People's Congress, which replaced Qaddafi's ruling Revolutionary Command Council, changed Libya's name from the Libyan Arab Republic to the Socialist People's Libyan Arab Jamahiriya. Qaddafi resigned as president to become the philosopher/ideologue of the revolution. Mixing populist ideological statements with political, social, and economic experimentation, Qaddafi undertook a cultural revolution. However, it was based not on the divine guidance of the Quran or the example of the Prophet, but on the iconoclastic thought of Qaddafi.

Libya's Al-Jamahiriya (state of the masses) was to be a people's state. Libya became a decentralized, participatory government of people's committees which controlled government offices, schools, the media, and many corporations. Qaddafi encouraged people's committees to take over Libya's embassies and mosques.

The economic policy of *The Green Book* included the abolition of private land ownership, wages, and rent in favor of worker control and participation in the means of production. Implementation of the new socialist experiment began in 1978. Libyans were limited to ownership of one home or apartment, so as to meet Libya's housing crisis. Private retail trade was all but abolished. Factory workers were encouraged to seize their factories, and self-management programs spread throughout Libya's public companies as workers overnight became partners in this workers' revolution.[9]

"Qaddafi's Islam"

While the Western media found it attractive to speak of Libya's Islamic fundamentalist state and to label much of what Qaddafi did or threatened to do as Islamic—Islamic laws, Islamic bombs, Islamic terrorism—many Muslims were increasingly incensed by Qaddafi. Qaddafi subordinated Islam to his interests and revolutionary ideology, used Islam to export his influence in the Muslim world, and earned the condemnation of Muslims inside and outside Libya for his unorthodox interpretation of Islam. Qaddafi, not the *ulama,* defined Islam. Though he used Islam to legitimate Arab socialism and his radical populist state, he did not tolerate alternative Islamic voices and visions. He denounced the *ulama* as "reactionaries," rejected their authoritative role as guardians of Islam, and sought to reform their centuries-old interpretation of Islam: "As the Muslims have strayed from Islam, a review is demanded. The [Libyan revolution] is a revolution rectifying Islam, presenting Islam correctly, purifying Islam of the reactionary practices which dressed it in retrograde clothing not its own."[10]

Qaddafi maintained the right of all learned Muslims (not just the traditional religious leadership) to interpret Islam. This provided the rationale for Qaddafi to freely and idiosyncratically reinterpret Islamic belief and practice. He produced what his critics regarded as a radical Islam that rejected key aspects of Islamic tradition to support Libya's populist state and his Third International Theory.

Contrary to stereotypes, Qaddafi's Libya was not an Islamic fundamentalist state tied to an inflexible literalist return to Islam. Qaddafi's brand of Islam was highly innovative and, for many Muslims, a radical revisionism tantamount to heresy. The major example of Qaddafi's innovative interpretation of Islam concerned the place of Islamic law, the Sharia, in the state. Traditionally, the accepted criterion for an Islamic state was a Muslim ruler's official acknowledgment of the Sharia. But Qaddafi set Islamic law aside. *The Green Book* replaced the traditional, comprehensive role of the Sharia. The Quran and *The Green Book,* which Qaddafi maintained was rooted in the Quran, became the ideological foundation of Libyan society. However, the Quran was restricted to private life (religious observances such as prayer, fasting, almsgiving) while Qaddafi's *Green Book,* not Islamic law, governed politics and society.

Qaddafi also boldly denied the authenticity and binding force of many prophetic traditions (Muhammad's exemplary practice), changed the dating of the Muslim calendar, declared that the pilgrimage to Mecca was not obligatory, equated the alms tithe (*zakat*) with social security and, contrary to tradition, maintained that this tithe could be a variable rate rather than a fixed (2.5 percent) tax.[11]

Qaddafi's usurpation of religious authority and his interpretation of Islam set him on a collision course with religious leaders who resented his attack on their authority and interests as well as his reinterpretation of Islamic doctrine. The *ulama* condemned Qaddafi's "innovative" interpretations—his rejection of the binding force of prophetic traditions, his substitution of *The Green Book* for Sharia governance, and his abolition of private property—for their deviation from Islamic tradition. Qaddafi justified his autocratic rule as a religiopolitical leader/ideologue and his rejection of traditional religious leaders in the name of the people's authority and Islamic social justice. People's popular committees were "instructed to 'seize the mosques' to rid them of 'paganist tendencies' and of *imams* [religious leaders] accused of 'propagating heretical tales elaborated over centuries of decadence and which distort the Islamic religion.' "[12]

Qaddafi's authority and legitimacy have also been challenged by Islamic movements such as the Muslim Brotherhood and the Islamic Liberation Organization (ILO). In a country where organized dissent is extremely dangerous, a regime which has used Islam to enhance its legitimacy has faced opposition in the name of Islam not only from the religious establishment but from Islamic movements. These activists share Qaddafi's criticism of the religious establishment as obstacles to change, but regard Qaddafi as a military opportunist who has manipulated and distorted Islam for his own ends. Qaddafi in return regards the Brotherhood, a strong religiopolitical movement in Egypt and the neighboring Sudan, as a political organization whose ideology, goal, and organization (antisocialist, Pan-Islamic, hierarchical) are a direct threat to his populist, socialist, Arab nationalist vision. The ILO, active in Jordan, Tunisia, and Egypt as well as Libya, infiltrated the military and was implicated in attempted coups in 1983 and in 1984.[13] Reports of militant Islamic opposition have persisted in the late eighties and nineties. Members of groups referred to as al-Jihad and Hizbullah were executed in 1987. Muslim student activism became more noticeable in 1989 with greater numbers of women wearing head scarves on campus, student demonstrations at Tripoli's al-Fatah University which were attributed to the Muslim Brotherhood, and a resumption of political discussions at mosques.[14] In the fall of 1989, Islamic militants participated in bloody confrontations with security forces and were denounced by Qaddafi as a "cancer, the black death and AIDS."[15] One observer has concluded:

"The Islamic movement in Libya may well form the best articulated and potentially most powerful opposition in the *Jamahiriyah.*"[16]

Foreign Policy

Since Qaddafi envisioned himself as the major revolutionary leader in the Muslim world, Islam served as an important element in Libya's export of its cultural revolution. Like Saudi Arabia, Libya under Qaddafi combined appeals to Islamic solidarity with oil wealth to promote its role as a leader in the Islamic world. Islam was never the only factor but one of many in Libya's foreign policy, joined with Arabism, Africanism, and a broader Third Worldism. Moreover, separating rhetoric from reality often proves difficult, given Qaddafi's pragmatism, his penchant for exaggerated rhetoric, and the Western media and the United States' bestowal on him of a larger-than-life image as a major source of worldwide terrorism.

Qaddafi's foreign adventurism grew out of his desire to fill the shoes of his hero Gamal Abdel Nasser as the leader of the Arab world and to be acknowledged as a leader and hero in Africa and the Third World. His militant universal vision drew its early inspiration from his interpretation of Islam, its expansionist universal message and mission, and its relevance to world liberation. He also learned from the example and methods of Egypt's Nasser and King Faisal of Saudi Arabia, who both had used Islamic rhetoric and organizations in the late sixties to compete for international leadership in the Muslim world. As a result, where feasible, Qaddafi combined missionary activity and propaganda with revolutionary politics. His radical politics proceeded from a worldview which demonized the West, especially the United States, and blamed the ills of the Arab and Muslim worlds (as indeed of the entire Third World) on European colonialism, American neocolonialism, and the state of Israel. Qaddafi rejected communism (Marxist atheism and its values) as an ideology, but had no difficulty aligning Libya with the Soviet Union and Eastern Europe.

For Qaddafi, the goal in a world dominated by the West was retribution and the struggle for national liberation.[17] He employed ideological appeals to Pan-Arabism, Pan-Islam, and Third World liberation. Libya funded governments, Islamic organizations, and extremist movements of many stripes around the world. Libya donated schools, mosques, and hospitals, and funded economic development projects as well as kidnappings, bombings, and murder. Qaddafi called for the destruction of Israel as easily as he advocated the overthrow of Saudi Arabia or Egypt.

He supported, claimed to support, or was claimed to have supported tyrants like Idi Amin of Uganda and Jean-Bédel Bokassa of the Central African Republic, Muslim minority liberation movements in the Philippines and southern Thailand, Islamic extremist movements from Egypt to Indonesia, Abu Nidal's radical Palestinian group, and the Irish Republican Army (IRA).

Because of the early identification of Libya as a "fundamentalist Islamic state," owing to its implementation of Islamic laws and criminal punishments, Libyan state-sponsored terrorism, regardless of its motives, came to be indiscriminately associated with Islam whether the actors were the Irish Republican Army, Arab or Palestinian organizations, Maoists, or Islamic extremists. Regardless of the orientation or authenticity of a movement, to be associated with Libya was often tantamount to admission of violent radicalism. Indeed, many governments in the Muslim world often found in Libya a ready-made ostensible cause for discrediting opposition movements.

The primary organizational example of Libyan use of Islam to promote its activities in the Muslim world is the Islamic Call Society. Within the post-1967 context of the Muslim world, with its renewed emphasis upon Islamic identity and solidarity domestically and internationally, Libya's Revolutionary Command Council established the Society in 1972 as a major vehicle of Libya's foreign policy. Its goal was not simply the preaching and promotion of Islam at home but its spread abroad as an integral part of Libyan diplomatic and political objectives. Mass media, religious publications, missionary activities, mosque building as well as medical and social welfare services were among the main tools for the spread of Libyan political influence. Recalling Libyan influence in Africa, a former Gambian diplomat noted:

> Those of us who have served as diplomats from Africa in these parts of the world know that significant amounts have been spent on these kinds of ventures. There is no Muslim country or country with a significant number of Muslims that has not received some financial support from the Libyans to build a mosque of some size.[18]

Combining diplomacy with the activities of the Islamic Call Society, Libya in the seventies and early eighties competed with Saudi Arabia in its use of Islam, employing its petrodollars to extend its influence from Muslim student groups to liberation movements. Thus for example, in the early seventies Libya was one of the first and the strongest supporters of the separatist Moro National Liberation Front (MNLF) in the southern Philippines, giving sanctuary to its leaders and extending aid for religious activities and refugees as well as military and diplomatic support.[19] Both Saudi Arabia and Libya were prime movers in bringing the Marcos government and the MNLF together for the ill-fated Tripoli Agreement.

The fixation of the Reagan administration on Qaddafi and Libya, and the attention that the media lavished on them, often elevated Qaddafi and his relatively small African state to a global significance which far exceeded his resources and capabilities and overshadowed the more pervasive terrorist activities of the Soviet Union and Syria. In addition, Libya's reputation as an Islamic fundamentalist state led to the identification of Qaddafi and terrorism with Islam, and the equation of all Islamic activists with radicalism and extremism. By the mideighties governments in the Middle East, Africa, and Southeast Asia as well as the United States and Europe were wary if not condemnatory of Libyan foreign intervention and violence. Libya's reputation as a radical regime that promoted revolution and terrorism was firmly established. For the Reagan administration, Qaddafi and Khomeini became symbols of worldwide terrorism, as menacing as the "evil empire." Many governments in Africa, a major area of Libyan activity, no longer regarded its presence and that of the Islamic Call Society as a welcome source of development funds and greater Islamic solidarity. In their eyes, Libya's political support for militant Islamic activists and revolutionary groups constituted a threat to their security far outweighing its aid.[20] Libya's influence and its effective export of Islam waned in the eighties in many parts of the world, as many governments and Islamic organizations became leery of Libya's tactics and as its oil revenues dried up and its relations with the Soviet Union cooled.

The Sudan

On September 8, 1983, Gaafar Muhammad Nimeiri declared an "Islamic revolution" that would impact Sudanese politics, law, and society. The Sudan, Africa's largest country, was henceforth to be an Islamic republic governed by Islamic law. Like Muammar Qaddafi, Nimeiri had come to power in a military coup in 1969. When the Free Officers, led by Colonel Nimeiri, seized power in their May revolution, three important ideological forces were at work in the Sudan: Islam, Gamal Abdel Nasser's Arab socialism, and communism. A failed communist coup in 1971, led by his former leftist allies, transformed Nimeiri into a committed anticommunist and, as a result, a close ally of the United States and Egypt's Anwar Sadat. The link to Sadat was so strong that the Sudan was the only major Arab country that did not break off diplomatic relations with Egypt after Sadat signed the Camp David Accords.

Personal as well as political factors influenced Nimeiri's espousal of an Islamic direction in his life and government. By 1977 domestic politics and the broader Islamic revivalist currents in the Arab world had come to serve as catalysts for the intensification of Islam's role in Sudanese politics. The appeal to Islam offered Nimeiri a new way out of a deteriorating situation. A series of events during the seventies had progressively narrowed his political options. After the abortive coup in 1971 in which he narrowly escaped death, he reportedly became more religiously observant, abstaining from alcohol and gambling, as well as resolutely antileftist. However, Nimeiri's own brand of Arab socialism had failed as a national ideology, proving unable to garner popular domestic support in the face of a collapsing economy, nor could it unify the disparate religious and ethnic–tribal interests. The Sudan's national debt spiraled out of control. Responding to pressures from the World Bank and the International Monetary Fund, the Sudan had lifted government subsidies on staples such as bread and sugar, causing popular antigovernment demonstrations and food riots in 1979 and again in 1982. Insurrection grew in the predominantly non-Muslim South. Nimeiri continued to be challenged by the National Front, an alliance of national Islamic organizations. He countered and preempted his Islamic critics in the National Front by himself harnessing religion to enhance his legitimacy. His public statements emphasized a holistic understanding of Islam, similar to that espoused by Islamic organizations that had opposed his rule, the Ansar or followers of the Mahdi and the Muslim Brotherhood. He even wrote a book, *Why the Islamic Way?*, about his increased emphasis upon Islam, advocating the application of Islamic law in the Sudan.[21] Nimeiri's appeal to Islam had continuity with the Islamic character of Sudanese political history and social culture, and thus, he believed, had the potential to consolidate popular support among Sudan's 70-percent Muslim population even if it threatened to alienate the Christian and animist southern third of the country.

Nimeiri struck a deal with the National Front in 1978 by signing the National Reconciliation Pact, which enabled his Islamic opposition to return to public life. Islamic political parties were able to compete successfully for parliamentary seats in national elections. The Muslim Brotherhood, an Islamic activist organization with a wide following among university graduates and professionals in urban areas, became closely associated with the government. Nimeiri appointed Dr. Hassan al-Turabi, the Sorbonne-educated leader of the Brotherhood and former dean of the University of Khartoum Law School, to the post of attorney

general. Muslim Brothers were to be found in the cabinet and in important positions in key institutions. Islam increasingly became a rallying
point as well as a source of division and opposition in Sudanese politics.

Nimeiri's "Islamic" laws of September 1983 marked the formal institutionalization of Sudan's Islamic path. Traditional Islamic legal punishments such as flogging for alcohol consumption, amputation for theft,
and death for apostasy were implemented. In contrast to Libya and
Pakistan, where Islamic punishments were legalized but seldom carried
out, amputations for theft became common in the Sudan. The Sudan's
Islamization program was very much "Nimeiri's Islam." New "Islamic"
laws were decreed on a weekly basis to suit his needs and whims. Islamic
regulations and policies were hastily formulated in a piecemeal and ad
hoc fashion, without consultation with the attorney general or chief
justice. They were issued by presidential decree, not by legislative action. Nimeiri circumvented the Sudan's duly established judiciary by
creating special "decisive justice courts." Thousands were arrested and
brought before government-appointed judges, whose courts often functioned more like military tribunals, employing "Islamic" punishments
such as flogging quite liberally for a variety of crimes. Government
officials and military leaders were required to take a pledge of allegiance
to Nimeiri as a Muslim ruler (a requirement imposed by the early caliphs
of Islam). Nimeiri even attempted unsuccessfully to take the title Imam,
that is, religiopolitical leader of the Muslim community.

In 1983–84 Nimeiri used flamboyant public acts to dramatize his new
Islamic order. He released thirteen thousand prisoners to give them a
"second chance" under Islamic law; poured eleven million dollars' worth
of alcohol into the Nile in a media event that gained both national and
international attention; and in May 1984 banned European-style dancing and subjected a nightclub owner to twenty-five lashes for permitting
mixed (heterosexual) dancing.[22]

In the socioeconomic sphere, new guidelines were enacted for taxation and banking. The Zakat Tax Act of 1984 replaced much of the
state's taxation system with Islam's alms tax. The Sudan now joined a
number of self-styled Islamic governments, like Pakistan and Iran, in
passing legislation that empowered the state to levy, collect, and distrib-
ute what had been a voluntary alms tithe. Nimeiri's stated intention to
convert all of the Sudan's banks into interest-free institutions was especially controversial. The move to an Islamic banking system in the Sudan, as in Pakistan and Iran, was to be the first step in basing the entire
economy on Islamic principles. These policies greatly disturbed many in
the Sudan as well as foreign interests, in particular the U.S.-based multinational companies operating there.

Nimeiri's Islamization program fragmented more than unified the Sudan, generating opposition among many sectors of society who regarded him as little more than an Islamic dictator. Initially, Islamization had proven popular in the Muslim North, where many felt that, as a result, crime and corruption were on the decrease; public floggings and amputations drew large supportive crowds. However, Nimeiri's use of Islam to expand his power and justify an increasingly repressive regime, erratic decisions by "decisive justice courts," the indiscriminate use of flogging, and the imposition of Islamic law on non-Muslims undermined his image at home and abroad. Sadiq al-Mahdi, a former prime minister and the Oxford-educated great-grandson of the Sudanese Mahdi, the Islamic revivalist reformer who had driven out the British and established an Islamic state in the Sudan in the 1880s, dismissed Nimeiri's "Islamic" September laws as opportunism. He argued that the imposition of Islamic law was premature and unjust, requiring first the creation of a more socially just society. For this, Sadiq was eventually imprisoned. Nimeiri's Islamization of society disrupted the fragile peace treaty achieved in the early seventies with the non-Muslim South and led to a new civil war, led by John Garang's Sudan People's Liberation Army (SPLA), the military wing of the Sudan People's Liberation Movement (SPLM).

Internationally, conservative Muslim states like Saudi Arabia, to whom Nimeiri looked for financial aid, became concerned about a negative image of Islam and Islamic justice in the Sudan, as the international media reported on a seemingly endless series of floggings and amputations. The United States and international organizations expressed concern over the violation of human rights caused by the imposition of Islamic law on non-Muslims. The U.S. Congress threatened to cut aid to the Sudan. A host of American officials and diplomats traveled to Khartoum to communicate U.S. concerns. The United States froze $114 million in economic aid in December 1984, siding with the IMF in pressing the Sudan to introduce economic reforms in order to control its soaring deficit of $9 billion.

At first, it seemed as though Nimeiri's "Islamic authoritarianism" was tempered. He responded to his international critics with a series of reforms which moderated his Islamization program. He discontinued the floggings and amputations in the North, backed away from his plan to divide the South, and gave assurances that Islamic law would not be implemented there. As he moderated his push for Islamization, talk of an imminent introduction of an interest-free Islamic economy also subsided.

However in 1985, amid growing domestic criticism of his Islamization program, Nimeiri desperately sought to silence his critics, rally popular

Muslim support, and find a scapegoat for the failures of his regime. In January, masking authoritarianism in the guise of protecting Islamic orthodoxy, on grounds of apostasy he arrested, tried, and executed the seventy-six-year-old founder and leader of the Republican Brothers, Mahmud Muhammad Taha. Many Muslims, including followers of Sadiq al-Mahdi, the Muslim Brotherhood, and the local Sufi leaders, had long regarded Taha's religious claims and his modernist reinterpretation of Islam as not simply liberal reformism but heresy. Taha had earned Nimeiri's wrath because of his opposition to Nimeiri's sectarian politics and attempts to implement Islamic law. Nimeiri's action seemed the move of a desperate man; his actions were seen by many not only as contradicting the Sudan's long tradition of political tolerance, but also as "a violation of Islamic law—as determined by an appeals court *ex-post-facto.*"[23]

In March 1985 Nimeiri continued his attempt to salvage his tottering regime, answer his critics, and redirect blame for the Sudan's ills away from himself. He moved to eliminate the Muslim Brotherhood as a political force, making it a scapegoat for the failures of his regime. In a statement that seemed to be meant more for his American allies than the Sudanese people, Nimeiri claimed to have thwarted a coup by the Muslim Brotherhood, who, he charged, were armed by Iran in order to overthrow his pro-American government. Nimeiri dismissed all members of the Muslim Brotherhood from the government and ordered the arrest of two hundred of its leaders, including Hassan Turabi.

The removal of Brotherhood leaders from key positions and new government "reforms" were a response to the United States, Egypt, and Saudi Arabia as much as to his domestic critics. Indeed, he took these steps immediately after the visit of Vice President George Bush and a U.S. delegation. The timing was such that, both within the Sudan and throughout the Arab world, rumors circulated that the four conditions Bush presented for lifting the freeze on American economic aid included discontinuation of Islamic criminal punishments (*hudud*) and dismissal of Islamic activists from the government and its institutions. The other two alleged points were halting contacts with Libya and accepting the economic reforms demanded by the IMF. The fact that Bush ended his visit by announcing resumption of U.S. aid, and that a team from the World Bank left for the Sudan the following day, appeared to confirm U.S. responsibility for Nimeiri's new initiatives. This perception was reinforced in late March when the Sudan's government, yielding to IMF and U.S. pressures, lifted subsidies on staples, an act which proved to be the undoing of the regime. As a result, Islamic activists in the Sudan and the broader Muslim world were convinced that a bias against Islam was behind U.S. policies—specifically, opposition to the implementation of

Islamic law and the involvement of activists in government. The State Department did little to counter these assertions. If government officials seemed worried about militant Islam, Islamic activists were convinced they were victims of the sons of the Crusades, a Christian West that still sought to dominate. However much both proceeded from stereotypes, their fears were grounded in their experience. For the United States, Islamic activists in government meant Islamic law and punishments, civil war, and infringement on minority rights. For activists, anti-Islamic America continued to intervene, to dictate policy and control the course of their lives from Washington. Each side regarded the other as a threat to its interests.

Critics of American policy accused the United States of hypocrisy in its concern about human rights in the Sudan. Many believed that, despite its democratic principles, the United States had been willing to tolerate autocratic rulers and dictators in the Sudan as in other Third World countries, making its support contingent solely upon a ruler's strong anticommunist position. Yet when Islam was involved, new criteria and the question of human rights were invoked to justify U.S. intervention in the policies of a sovereign country despite U.S. principles of self-determination. Anger at such intervention in the economic and political life of the Sudan became evident in the days leading up to the fall of Nimeiri, as anti-Americanism surfaced in antigovernment demonstrations and marches. Perhaps fittingly, it was during Nimeiri's visit to the United States that his government fell from power.

Islam, Democracy, and the Return to Military Rule

Nimeiri was overthrown on April 5, 1985, by a military coup. Democracy was restored by the interim military government in 1986, when a civilian government was elected with Sadiq al-Mahdi as prime minister, and parliament was reconstituted. The role of Islam and the Sharia continued to be an issue in Sudanese politics. While Sadiq distanced himself from the excesses of Nimeiri's Islamization program, he was never able to resolve the issue of Islamic law, which remained a bone of contention between the rebels in the South and the government in the North. Moreover, Sadiq al-Mahdi took a more independent position toward the United States, holding meetings with Iran and Libya. After three years of democracy, his ineffectual government was replaced by the military regime of Gen. Omar al-Bashir, who seized power in June 1989.

In the immediate aftermath of this coup, Bashir's military government suspended parliament and all political parties; imprisoned leaders of the major religiopolitical parties, including Sadiq al-Mahdi and Hassan

Turabi; and pledged itself to restore political stability, eliminate sectarian politics, and freeze the implementation of Islamic law. However, the issue of Islamic law again proved an insurmountable obstacle. The SPLM rejected Bashir's decision to resolve the question of implementation of Islamic law through a national referendum, believing that the northern (Muslim) majority would thus be put in the position of seeming to vote against Islam.[24] An attempt by former president Jimmy Carter in late 1989 to negotiate a settlement between the Bashir government and the SPLM also failed in large part over the issue of Islamic law.

Despite its initial disclaimers, the regime proved to be strongly influenced—ideologically, if not organizationally—by the National Islamic Front (NIF) or Muslim Brotherhood.[25] NIF members and sympathizers hold important positions in government, contributing to the further polarization of Sudanese society. Despite SPLM insistence on a secular state, Bashir on December 31, 1990, promised to implement Islamic law. Stories of interrogation and even torture by the NIF of its opponents abound. At the same time Bashir has been quick to crush any and all opposition to the military from professional associations. The seriousness of the crisis is reflected in the observation of a longtime scholar on the Sudan:

> The situation in the Sudan is one of multiple crises. The heritage of ineffective rule, both civilian and military, is frightening. The issue may in fact have changed from who will rule Sudan to whether or not Sudan will be able to survive in any meaningful fashion. . . . The military leaders of Sudan will have to be more flexible so that Sudan will not repeat the old cycle again or descend into anarchy.[26]

In the past two decades the people of the Sudan have lived through two military dictatorships and a democratically elected government. Government and opposition parties, autocracy and democracy have been legitimated in the name of Islam.

Egypt

> I am Khalid Islambuli, I have killed Pharaoh and I do not fear death.
>
> —Lt. Khalid Islambuli

Such were the words of the assassin who had shot and killed Pres. Anwar Sadat in an act which stunned the world and, in the wake of the Iranian Revolution, dramatically drew world attention to radical Islam in Egypt. Like Iran, Egypt had long been regarded as among the most modern Muslim countries, a leader in the Arab world politically, militarily, culturally, and religiously. It had offered a barometer for modernization which was predominantly Western and secular in orientation. The seemingly sudden eruption of militant Islam contrasted sharply with modern expectations and values. Today Egypt provides a remarkable example of the diverse and complex impact of Islam on sociopolitical development. For several decades Islam has been part of the political scene in Egypt, used by both the government and its opposition. Egypt has been a cradle of both Arab nationalism and Islamic revivalism under its last three rulers: Gamal Abdel Nasser, Anwar Sadat, and Hosni Mubarak.

Sadat's Islamic Path

Gamal Abdel Nasser's death in 1970 shocked Egypt and the Arab world. Millions of Egyptians poured into the streets of Cairo for his funeral. Although the pride fostered by Nasser and the credibility of Arab socialism had been shattered by the catastrophic Arab defeat in 1967, the death of the charismatic Nasser left a void that few men could have filled. Anwar Sadat, Nasser's successor, had been completely eclipsed by Nasser. Sadat, who was neither a charismatic nor a popular leader, lacked political legitimacy. Tellingly, in the early days of his rule, Sadat's picture was always seen alongside that of Nasser. Struggling to be regarded as a leader in his own right and to enhance his political legitimacy, he turned to and relied heavily upon Islam. Sadat appropriated the title "The Believer President," had the mass media cover his praying at the mosque, increased Islamic programming in the media and Islamic courses in schools, built mosques, and employed Islamic rhetoric in his public statements. Sufi brotherhoods and the Muslim Brotherhood, suppressed by Nasser, were permitted to function publicly. The creation and growth of Islamic student organizations on campuses was promoted to counter the opposition of Nasserites and leftists who opposed Sadat's pro-Western political and economic policies. Sadat used Islam to legitimate key government actions and policies such as the 1973 Egyptian–Israeli war, the Camp David Accords, and Muslim family-law reforms, and to denounce as extremism the increasing challenges to the regime by Islamic activists during the late seventies.

By 1977, Islamic organizations such as the Muslim Brotherhood and student groups were securely established. They became more independent and critical of Sadat's policies: his support for the Shah of Iran and condemnation of the Iranian revolution as a "crime against Islam,"[27] the

Camp David Accords, pro-Western economic and political ties, and the failure of his government to implement Islamic law. More ominously, a new crop of secret revolutionary groups, some founded by radicalized former Muslim Brothers imprisoned under Nasser, began to challenge what they regarded as Sadat's hypocritical manipulation of Islam and to reject the moderate posture of the Muslim Brotherhood. Militant groups like Muhammad's Youth, the Army of God, and the Islamic Society resorted to acts of violence in an attempt to overthrow the government.

Sadat was seen by his critics as a prime example of Egypt's Westernized elite in both his personal and his political life. His penchant for imported suits and pipes as well as the high public image and international profile of his half-British wife Jehan, which sharply contrasted with the more reserved public image of wives of Egyptian rulers and politicians, offended his Islamic critics. Many activists, moderate as well as radical, rejected Sadat's reform of Muslim family law (the law governing marriage, divorce, and inheritance) as "Jehan's law," meaning that it was influenced by her Westernized outlook. His open-door economic policy—resulting in the higher profile of America's presence, as symbolized by a towering new embassy and an influx of U.S. businessmen—along with his support for the Shah and criticism of Khomeini, and the Camp David Accords, were all regarded as evidence of Sadat's capitulation to the West. These critics condemned his political and military dependence on the West and the more pernicious threat, Western cultural penetration and acculturation, a theme which struck a responsive chord among many: "Culturally, fathers have seen their sons and daughters influenced by what they consider to be immoral Western behavior and believe them to be neglecting their traditional values. Thus now, as in the 1930's, there is an outcry against the cinema, theater, magazines, books, clubs, and all associations that popularize modern concepts such as individualism and the lack of concern with family problems and wishes of parents and older brothers."[28] Attacks against bars, nightclubs, cinemas, Western tourist hotels, and government institutions and personnel signaled their grievances. The militants believed that the liberation of Egyptian society required that all true Muslims undertake an armed struggle or holy war against a regime which they regarded as oppressive, anti-Islamic, and a puppet of the West.

Criticism and condemnation of Sadat seemed confirmed by his responses to his growing opposition. In February 1979 he called for the separation of religion and politics, a position seen as un-Islamic by Muslim organizations calling for an Islamic state and the implementation of Islamic law. Organizations of Islamic university students were banned.

Sadat used government control of nationalized mosques and religious institutions, and the dependence of the religious establishment on the government for its salaries, to dictate and control sermons and mosque activities. At the same time he attempted to extend state control to Egypt's more numerous private mosques, whose preachers were far more independent and increasingly critical of the regime. In April 1980 Sadat moved to blunt his critics with a highly publicized amendment to the constitution declaring that "Islam is the religion of the state" and that the Sharia was "the main source of legislation." Yet for radicals, this was mere window dressing: Sadat's track record, his failure to actually implement Islamic law, and his pro-Western policies rendered him an apostate who merited the death penalty. Sadat's growing authoritarianism and suppression of dissent, whether secular or religious, reached its apogee in 1981, when he imprisoned more than fifteen hundred people—Islamic activists, lawyers, doctors, journalists, university professors, political opponents, and ex-government ministers.

On October 6, 1981, Sadat was assassinated by members of the Organization for Holy War (Jamaat al-Jihad) as he reviewed a parade commemorating the 1973 war. Lt. Khalid Islambuli, the leader of the assassins, cried out: "I am Khalid Islambuli, I have killed Pharaoh and I do not fear death." Sadat's popularity in the West and his international image as an enlightened leader stood in sharp contrast to his growing unpopularity at home, where both secular and religious opposition referred to him as "Pharaoh." For many in Egypt, "Khalid therefore appeared as a sort of 'right arm' of the popular will, and not merely as a militant exponent of an Islamicist group."[29] Al-Jihad had been formed by remnants of another militant group, Muhammad's Youth or the Islamic Liberation Organization (ILO). ILO leaders had been executed for their 1974 attempt to seize the Technical Military Academy in Cairo as part of a foiled coup d'état.

The leadership and members of al-Jihad were drawn from a cross section of society: civilian, military, and religious. They included members of the presidential guard, military intelligence, civil servants, radio and television workers, and university students and professors. The rationale for their actions was articulated in a tract entitled *The Neglected Obligation*. Written by their ideologue, an electrician named Muhammad al-Farag, it maintained that jihad was the sixth pillar of Islam and that armed struggle or revolt was an imperative for all true Muslims so as to rectify the ills of a decadent society: "[W]e have to establish the Rule of God's Religion in our own country first, and to make the Word of God supreme. . . . There is no doubt that the first battlefield for jihad is the extermination of these infidel leaders and to replace them by a complete Islamic Order. From here we should start."[30]

For Sadat, as for Nasser, Islam was an instrument of foreign policy which did not determine but was determined by other political and economic priorities. He emphasized both Arabism and Islam and at times seemed to equate the two. However, reflecting his pro-Western policies, Sadat, unlike Nasser, did not emphasize anti-imperialism. Instead he increasingly emphasized the Islamic community (*ummah*) and the common interests of Muslim states, as well as "the commonality of interests between the Islamic states and the West; oil in exchange for technology and the common interest against the Soviet threat."[31] However, as previously noted, his pro-Western orientation and policies—economic reforms, the Camp David Accords, and support for the Shah of Iran—undermined his credibility with many Egyptians and alienated Islamic groups.

Hosni Mubarak and the Institutionalization of Islamic Revivalism

The style of both the Egyptian presidency and Islamic revivalism changed in the eighties after the death of Sadat. While Islamic revivalism in Egypt seemed to degenerate into confrontation and violence during the seventies, the eighties witnessed the growth of a broader-based, quiet revolution, overshadowed or eclipsed at times by the violence of a more vocal and militant minority.

In contrast to his predecessor, Hosni Mubarak pursued a path of greater political liberalization and tolerance, while at the same time responding firmly to those who resorted to violence to challenge the authority of the government. He distinguished more carefully between religious and political dissent and direct threats to the state. He crushed outbursts and riots fomented by Islamic militants, intervened in clashes between Muslims and Coptic Christians, and tried and executed the assassins of Sadat. At the same time he surprised many when he released more than 174 defendants in the Sadat trials acquitted by the courts. Unlike Sadat, Mubarak permitted religious critics public outlets for their opposition: they could compete in parliamentary elections, publish newspapers, voice criticism in the media. The government sponsored television debates between Islamic militants and representatives of the religious establishment, religious scholars from al-Azhar University. Government-run television and newspapers regularly featured religious programs and columns which were often independent in their tone and criticisms. At the same time, however, the government used members of the state-supported religious establishment to debate and dismiss militants as deficient in their knowledge of Islam and misguided, if not heretical, in their beliefs.

However, by the mid-eighties Mubarak's flexible response to Islamic activism had failed to coopt or silence his Islamic opposition. Political liberalization opened up new opportunities for the Muslim Brotherhood. The Brotherhood had become more of a major political, social, and economic force; it emerged as the chief political opposition in parliamentary elections and an effective social and economic actor through its publishing, banking, and social services. Muslim Brothers and other activists were dominant voices in professional organizations and syndicates of lawyers, doctors, engineers, and journalists. Increasingly, moderate activists such as the Muslim Brotherhood presented their criticisms and demands within the context of a call for greater democratization, political representation, and respect for human rights.

Mubarak's attempt to discredit and isolate religious extremists infuriated Islamic radicals. Radical groups became more assertive in Cairo, Alexandria, Asyut, and Minya. Muslim militants attacked Coptic Christian churches and destroyed their shops and property. Bars, nightclubs, cinemas, and video stores, symbols of Western influence and immorality, were burned or bombed in major cities and towns. Street demonstrations and protests were organized to demand the imposition of Islamic law. Popular militant clerics like Shaykh Hafiz Salaama led public demonstrations. In an attempt to control the spread of radicalism, in July 1985 the Mubarak government placed all private mosques under the Ministry of Religious Endowments, arrested Salaama, who had led demonstrations calling for the implementation of Islamic law, and closed down his mosque, a center of dissent.

The degree of discontent and polarization between the government and moderate opposition groups was reflected in the failure of opposition leaders to distance themselves completely from extremist acts. When a disturbed border policeman killed several Israeli tourists in the Sinai in December 1985, Muslim Brotherhood leaders, joined by opposition newspapers and political parties, described the killings as an act against the "enemies of the nation." Their ire was directed at both Israel and the United States. Ibrahim Shukry, leader of the Socialist Labor Party, reflected the mood of many: "This young man who has removed the shame from Egypt after Israel has bombarded the P.L.O. headquarters in Tunisia and after the Americans have hijacked the Egyptian plane [referring to the *Achille Lauro* incident]."[32]

In Asyut, Minya, Cairo, and Alexandria, Islamic student organizations once again dominated university student unions and pressured authorities for the implementation of Islamic law, gender segregation in classes, curriculum reform, restriction of mixed socials, and the banning of Western music and concerts. Their growth was fed by the government's inability to address continued chronic socioeconomic realities which had a disastrous effect upon the more than half of Egypt's fifty million citizens below the age of twenty. Hundreds of thousands of university graduates found jobs and housing impossible to obtain. Young couples often lived with their families or delayed marriage for years until they could find adequate housing.

The military also continued to prove vulnerable to infiltration by Islamic militants. In December 1986 thirty-three activists—including four military officers allegedly connected to al-Jihad, Sadat's assassins—were arrested and charged with plotting to wage a holy war in order to overthrow the government. The government continued quietly to purge those suspected of being "fundamentalists." Clandestine groups kept on growing and sporadically confronted the government, which increasingly struck back, sometimes indiscriminately. In 1989 as many as ten thousand Islamic militants were arrested. Thousands were held without charge; the Arab Human Rights Organization accused the government of routine torture.[33]

Mubarak's lackluster image, his failure to provide dynamic and creative political leadership, and a weak economy, worsened by the loss of jobs and foreign revenue from millions of expatriate Egyptian workers in the Gulf, led to a more diffident political liberalization program. At the same time it made his government more vulnerable to its Islamic critics, who countered that less economic and political dependence on the West, a rejection of the Camp David Accords, and the implementation of Islamic law offered the only solution.

Islamic activists, whether moderate or radical, failed to offer a specific concrete alternative program and were content simply to criticize the failures of the incumbent government. Both the nature and diverse agendas of these movements as well as the realities of Egyptian politics were causes. Many Islamic activists and organizations chose to emphasize the creation of a more Islamically oriented society through sociomoral rather than political reform. These activists focused on religiosocial programs and issues, rather than defining the nature of an Islamic state and its institutions. Even the Muslim Brotherhood emphasized building its base of support through educational and socioeconomic programs. In oppositional politics it was content to seek the implementation of Islamic law rather than directly challenging the head of state. For its part, the government simply criticized the Brotherhood's lack of specificity and maintained that, if it came to power, it would lack specific programs and so be doomed to failure. Yet the government, like most Muslim governments, has chosen not to test its conviction. It has neither allowed the Brotherhood to become a political party nor fostered political reforms which would allow a real electoral challenge. Indeed, whenever possible it has sought to coopt, control, or suppress any viable opposition.

The most important characteristic of Islamic revivalism in Egypt in the nineties is the extent to which revivalism has become part and parcel of moderate mainstream life and society, rather than a marginal phenomenon limited to small groups or organizations. No longer restricted to the lower or lower middle class, renewed awareness and concern about leading a more Islamically informed way of life can also be found among the middle and upper class, educated and uneducated, peasants and professionals, young and old, women and men. They are active in Quran study groups (run by both women and men), Sufi gatherings, mosques, and private associations. As a result, Islamic identity is expressed not only in formal religious practices but also in the social services offered by psychiatric and drug rehabilitation centers, dental clinics, day-care centers, legal-aid societies, and organizations which provide subsidized housing and food distribution or run banks and investment houses.

The *ulama* and the mosques have also taken on a more prominent role. The popularity of preachers such as Shaykh Muhammad Mitwali al-Shaarawi and Abd al-Hamid Kishk, an outspoken critic of both Nasser and Sadat, has made them religious media stars. They appear regularly on television and have newspaper columns. Their audiocassettes, pamphlets, and books are sold not in specialty shops but in ordinary bookstores, at airports and hotels, and by street vendors. The popularity of these preachers extends beyond Egypt to much of the Arab world. Thus their voices are heard not only in mosques or religious gatherings but also on cassettes played in taxicabs, on the streets, in shops, and in the homes of the poor and middle class alike. Wherever one finds magazines, audio tapes, or books being sold, there, amid the coverage of politicians and movie stars, will also be popular religious materials. Private, nongovernment mosques and their imams have for some time been a source of independent criticism and opposition to the government. While the laity (non-*ulama*), through modern Islamic organizations such as the Muslim Brotherhood, have exercised leadership in relating Islam to the demands and needs of modern society, more and more mosques and their religious officials have also become increasingly involved in providing much-needed social services. As a result Islamic revolutionaries, though active from time to time, have become more and more marginalized and limited in their appeal.

The Islamic Republic of Iran

The cry that comes from the heart of the believer overcomes everything, even the White House. . . . This wave has already spread
throughout the world, and the world is now liberating itself from the
oppression to which it has been subjected.

—AYATOLLAH KHOMEINI

For more than a decade Iran represented the embodiment of the Islamic
threat, and the Ayatollah Khomeini served as the living symbol of revolutionary Islam. If Khomeini denounced the West—and the United States
in particular—as the great Satan, for many in the West he represented a
medieval cleric, a menace to the Middle East and the West. From presidential statements to popular music and billboards, Khomeini soon became the man many Americans loved to hate.

For a West convinced of the stability of the Shah of Iran, the fall of the
Shah was unthinkable, a shock intensified by the victors, the mullahs of
Iran. Muhammad Reza Pahlavi had ruled for more than thirty years
(1941–79). Like Anwar Sadat, he had come to be viewed in the West as
an enlightened ally, the head of a modern state. The Pahlavis, like the
Sadats, spoke English, dressed in well-tailored Western clothes, and
appeared on American television, interviewed by the likes of Barbara
Walters. (They were like "us.") Yet in 1971 the world witnessed Iran's
celebration of the twenty-five-hundredth anniversary of Persian monarchy. Before an audience of the world's leaders, including the vice president of the United States, the Shah crowned himself emperor of Iran.
Under the Shah, oil revenues had been utilized in an ambitious modernization program, the White Revolution, whose goal was the modern
transformation of Iran by the twenty-first century. Iran possessed the
best-equipped military in the Middle East and enjoyed close relations
with the United States and Europe. That such a shah would be overthrown was unthinkable; that the Pahlavi dynasty would come crashing
down at the hands of a revolution led by a bearded, exiled ayatollah and
conducted in the name of Islam was incomprehensible.

Islam, Nationalism, and the State

Despite the popular Western image of Shii Islam as a religion of revolution and martyrdom, its relationship to the state in Iran throughout
Islamic history has been diverse and multifaceted. Islam in Iran incorpo-
rates a variety of outlooks and orientations. It has been capable of

multiple levels of discourse and interpretation. Indeed, once the 1978–79 revolution had dethroned the Shah and the smoke had cleared, the diversity within Iran's Shii community would surface in contention and conflict.

Religion and the state had been intertwined in Iran ever since the establishment of the Safavid dynasty (1501–1732), when Shii Islam was declared Iran's state religion. Iran's shahs claimed to rule in the absence of the Twelfth Imam, who had disappeared in 874 and whom devout Shii believed would return in a future age to end tyranny and usher in a new age of justice. The clergy often proved far from revolutionary. They accepted the necessity of temporal rule in fact, if not in theory or doctrine. The relationship of the *ulama* to the state in Iranian history varied from royal patronage to opposition, depending on the sociopolitical context. Their authority was often subordinated to and limited by rulers. At other times they led or participated in opposition movements. While the Safavids controlled the clergy, their successors, the less powerful Qajar dynasty (1794–1925), faced a more independent and assertive religious establishment who were not afraid to take to the streets in opposition.

On two more recent occasions—the Tobacco Protest of 1891–92 and the Constitutional Revolution of 1905–11—Islam and the clergy played an important oppositional role in the emergence of modern Iranian nationalism. Having previously thwarted attempts by the government to sell concessions to Europeans for the development of banking, railroads, and mining, religious and secular leaders joined together in protest movements first to protect Iranian national interests and then to constitutionally limit the Shah's power. The Tobacco Protest was a response to Nasir al-Din Shah's attempt to sell the tobacco concession to a British company (and thus create a monopoly). A religious proclamation (*fatwa*) legitimated a nationwide boycott led by religious leaders and merchants. The Constitutional Revolution utilized religious symbols, clergy, and mosques, which served as centers of political organization, to mount a popular protest movement through which Iranian nationalists sought to curb or constitutionally limit the absolutism and abuses of the monarchy.

Under Pahlavi rule (1925–78), religion was carefully controlled and the *ulama* remained relatively quiescent and apolitical. Both Reza Shah Pahlavi (1925–41) and his son, Muhammad Reza Shah Pahlavi (1941–78), controlled religion, combining cooptation with coercion of the religious establishment. For their part, Iran's religious leaders preferred quietism to political action, order to the threat of civil disorder, limited patronage to persecution. The creation of a modern secular school system and Western-based legal codes, as well as the control of many Islamic institutions by government ministries, sharply curtailed the power of the *ulama*. Their replacement by modern-educated judges, lawyers, and civil servants eroded the sources of their status and revenue.

The broader social impact of Muhammad Reza Shah Pahlavi's modernization programs, the White Revolution, which like his educational and legal reforms were influenced by the West, proved a mixed blessing. Despite accomplishments in education, health, and agricultural reform, the benefits of modern reforms went disproportionately to a small and growing group of modern, urban elites. The lights and glitter of modernized cities obscured the actual conditions of the urban poor and of the masses in the villages of Iran. While a minority prospered, a once agriculturally self-sufficient country was spending more than one billion dollars on imports. Those who poured into the cities from the villages expecting a better life, lacking requisite job skills, became unemployed inhabitants of overcrowded, congested urban slums: "For these millions, most of whom had been forced out of the villages into new shantytowns, the oil boom did not end poverty; it merely modernized it."[34]

Both the traditional merchant classes (*bazaari*) and the religious classes suffered as a result of the Western-oriented Pahlavi modernization program, which affected their lives from dress, education, and law to land reform and commerce. The *bazaari*, like the *ulama*, found Iran's tilt toward and dependence upon the West a threat to their status, economic interest, and religiocultural values. Reza Shah's dress codes in the twenties and thirties, which had mandated Western dress for men, banned the veil, and restricted the wearing of clerical attire, were now followed by the seeming Westernization of Iran's modern elites and many of its urban centers under his son. The merchants' wealth and power were threatened by the influx of Western banks and corporations and the new entrepreneurial class that emerged and prospered with state support.

Iran and the West

The West had long been a challenge and a threat to Iran. Although Iran had never been directly ruled by colonial powers, the Soviet Union in the north and Britain in the south had vied for influence. The danger of foreign intervention and dependence, symbolized by events provoking the Tobacco Protest, was realized in 1941, when Britain and the Soviet Union forced Reza Shah to abdicate in favor of his son. It was even more evident in 1953, when the Shah was driven into exile by a nationalist movement headed by Prime Minister Muhammad Mossadegh, whose nationalization of Iran's oil threatened the interests of Western oil companies. The Shah's return from Rome to Teheran, aboard an American military plane with the head of the CIA at his side, was orchestrated by the United States with British support. Iran's political, military, and economic ties with the West, and with the United States in particular, increased significantly. The United States, Britain, and France benefited from lucrative arms sales and helped train Iran's military and secret police (SAVAK). At a time when the United States was heavily committed in Vietnam and Britain was withdrawing its forces from the Persian Gulf, the Shah's Iran represented policies and interests that coincided with those of the United States, from the Shah's rejection of Nasserism and pragmatic relations with Israel, and his nation's stable presence in the Gulf, to its oil wealth and market for American products. American and European bankers and businessmen along with diplomats and military advisers enjoyed a high-profile presence in Iran. By the late sixties "pro-Pahlavi policy and support came to dominate the very highest levels of the American foreign policy establishment."[35]

Dissent began to grow and spread to a broader social base during the early seventies. Concern about foreign intervention and dependence on the West was not restricted to the traditional classes but included a generation of modern-educated, politically sophisticated, and socially minded Iranians. Middle-class nationalists and intellectuals joined merchants and clergy to voice their concerns for Iran's national identity and independence. Calls for social and political reforms increased amid complaints about concentration of wealth, corruption, growing political repression, and excessive dependence upon the West.

Repression and Dissent

By the seventies the Shah was no longer an inexperienced youth placed on the peacock throne by foreign powers, but an entrenched, increasingly autocratic ruler. As James Bill has commented, his policy of "cooptation with coercion, repression with reform," gave way to "increasing megalomania."[36] The October 1971 celebration of the twenty-five-hundredth anniversary of Persian monarchy was a symbolic turning point, marking at once the acme and the incipient decline of Pahlavi fortunes. More than two hundred million dollars was spent to gather dignitaries from around the world at Persepolis, the uninhabited pre-Islamic capital of Persia, where the Shahanshah (Shah of Shahs) aligned himself with the ancient Persian king, Cyrus the Great, the King of Kings. The week-long celebration, featuring a feast catered by Maxim's of Paris that included twenty-five thousand bottles of wine, epitomized the Shah's insensitivity to his critics and to Iran's Islamic identity and traditions. Royal extravagance and excess were combined with a Western-style celebration at a pre-Islamic site. Thereafter, the bloody suppression of student demonstrations prefigured a new policy which tolerated little dissent and "resulted in a reign of terror."[37]

Iranian reformers, the government's opposition, represented a cross section of society: nationalists and leftists, secularists and religious, traditional elites (merchants and clergy) and modern elites. Critics of Iran's military, economic, and political dependence on the West also feared the cultural alienation that would result from the progressive Westernization of Iranian education and society. Issues of national identity, Western influence, social justice, and political participation cut across social and ideological boundaries, and were articulated by lay and clerical critics alike. Among the most influential were Jalal-e-Ahmad, Mehdi Bazargan, Dr. Ali Shariati, and the Ayatollah Khomeini. Their ideas and leadership influenced a generation of students, intellectuals, and professionals (scientists, engineers, journalists) drawn from both the traditional and the modern middle class. Islamically oriented students and professionals would join with the clergy, seminarians, and *bazaari* in the broad-based opposition movement that toppled the Shah.

Jalal-e-Ahmad (1923–69) a secularist and socialist, gave voice to the fears of many that modern educational and social reforms were resulting in a seductive process of cultural assimilation, "Weststruckness," which threatened to rob Iran and particularly the younger generation of its sense of national identity: "We're like a nation alienated from itself, in our clothing and in our homes, our food and our literature, our publications, and most dangerously of all, our education. We effect Western training, we effect Western thinking, and we follow Western procedures to solve every problem."[38]

Jalal-e-Ahmad did not see a simplistic world of monolithic, stereotypical choices, either a path of modernization/Westernization of society or a retreat to the past. He opted for a third alternative, a return to Irano-Islamic culture as a source of national identity, unity, history, and values. His secular intellectual indictment of Westernization, with the call for continuity of modern Iran's identity and culture with its past, was a primary agenda for others as well, religiously oriented lay and clerical ideologists, laymen like Mehdi Bazargan and Ali Shariati and clerics like the Ayatollah Khomeini.

In 1962, the year when Jalal-e-Ahmad's *Gharbzadegi,* a criticism of "Weststruckness," was published, Mehdi Bazargan (b. 1907) delivered a lecture entitled "The Boundary Between Religion and Social Affairs" on the relationship of religion to politics. Bazargan, a French-trained engineer with a strong Islamic commitment, had been jailed in 1939 for his opposition to Reza Shah and again from 1962 to 1969 for his opposition to Muhammad Reza Shah. He had been active in Prime Minister Mossadegh's National Front and cofounded the Liberation Movement of Iran to bridge the gap between modern secular and traditional religious Iranians and work toward a more Islamic state and society. In 1979, with Khomeini's approval he would become the provisional prime minister of the Islamic Republic of Iran.

Bazargan's political activism was inspired and shaped by his Islamic orientation. He combined a traditional religious outlook and vocabulary with modern concerns. Thus he was effective with modern professionals and also, because he spoke their language, with the *ulama*. He had long argued that because of the interrelatedness of religion and politics in Islam, the *ulama* should move away from their political neutrality and become actively involved in politics to bring about the renewal of Islamic society in Iran.

The socialist inclinations of Jalal-e-Ahmad and the Islamic reformist spirit of Bazargan were combined in Dr. Ali Shariati (1933–77), whose revolutionary populist Islam was at once appealing to secularists and leftists, and in particular to university students, who often found the traditionalism of their religious leaders unconvincing and the Western secular outlook of many professors disorienting. Shariati incorporated many of the reformist currents of his times: opposition to the Shah, rejection of Westernization, religious revivalism and social reform.

The son of a progressive preacher-scholar, Shariati had earned a doctorate from the Sorbonne. After his return to Iran in 1965, Shariati was the subject of both controversy and adulation. Initially imprisoned for his opposition to the Shah during his student days in Paris, he proved an enormously attractive and successful lecturer in Teheran, drawing thousands of admiring students, and developed a wide following among students, intellectuals, and the Left. More than one hundred thousand copies of his lectures and writings were published and distributed.[39] Regarding Shariati increasingly as a threat, the government denounced him as an "Islamic Marxist," imprisoned him in 1973, and finally permitted him to leave Iran. Dying shortly after his arrival in England of an apparent heart attack (his supporters suspected SAVAK of foul play), Shariati became even more influential in death than in life. His talks were collected, printed, and distributed widely. He became a hero and ideologue of the revolution in Iran, and after that revolution his writings and ideas would spread throughout much of the Muslim world, finding attentive audiences from Cairo to Jakarta.

Shariati preached what may be described as an Islamic theology of

Third World liberation, an indigenous populist Shii ideology for sociopolitical reform. It combined a reinterpretation of Islamic belief that incorporated modern sociological language (Emile Durkheim and Max Weber) with the Third World socialist outlook of Frantz Fanon and Che Guevara, whom he had come to admire during his student days in Paris, which had coincided with the Algerian and Cuban revolutions. However, in contrast to Fanon's rejection of traditional religions, Shariati, like Jalal-e-Ahmad, insisted that the defeat of Western imperialism required the reclamation of Iran's Islamic roots, its national, religious, and cultural identity. Shariati emphasized the dynamic, progressive, scientific nature of Islam and the need for a thoroughgoing reinterpretation of Islam to reclaim this forgotten heritage, reverse the retrogressive state of Islam, and revitalize the Muslim community. An innovative Islamic reformer, he was often at odds with both the religious traditionalism and political quietism of many *ulama* and the Westernized, secular outlook of many university professors. Both the political and the religious establishments were wary of him.

Like Jalal-e-Ahmad, Shariati denounced the evils of "Weststruckness": "Come, friends, let us abandon Europe; let us cease this nauseating, apish imitation of Europe. Let us leave behind this Europe that always speaks of humanity, but destroys human beings wherever it finds them."[40] Shariati combined denunciation of Western imperialism with an Islamic brand of socialism. The goal was national identity and unity and socioeconomic justice for a politically subjugated and economically exploited Iran, caught in the grips of "world imperialism, including multinational corporations and cultural imperialism, racism, class exploitation, class oppression, class inequality, and *gharbzadegi* [Weststruckness]."[41]

The decline of Muslim society was due not only to Western imperialism but to a religious establishment which had allowed the dynamic, revolutionary ideology of "original" Shiism to become an establishment religion. Like numerous contemporary Islamic reformers, Shariati blamed the *ulama* for many of the ills of Muslim society. In the hands of the religious establishment, Shii Islam had become scholastic, institutionalized, and historically coopted by Iran's rulers, eclipsing the true, dynamic, pristine revolutionary religiosocial message of its early years. Traditional Islam under the *ulama* had become mired in a fossilized past of scholastic manuals, the popular, fatalistic opiate of the masses. It had ceased to be a social force, effectively addressing the changing realities of society. Shariati's "return to Islam" was not a retreat to the medieval Islamic worldview of conservative *ulama,* but a revolutionary vision of early Shii Islam which provided the inspirational basis for a modern reinterpretation of Islam. This would require men of vision—not the *ulama* but a religiously minded lay intelligentsia, Islamically oriented but with a knowledge and command of modern thought and methods.

For Shariati, true Islamic society was to be classless, a haven for the disinherited who were "plundered, tortured, hungry, oppressed and discriminated against." His was a populist message of a God who "had promised the wretched masses they would become leaders of mankind; He had promised the disinherited they would inherit the earth from the mighty. . . . [T]his universal Revolution and final victory is the conclusion of the one great continual justice-seeking movement of revolt against oppression."[42] It remained for the Ayatollah Khomeini to lead that revolt to its triumphant if not successful conclusion.

If Shariati was the ideologue of the Iranian revolution, the Ayatollah Khomeini (1902–89) was its living symbol and architect. Although he was a junior ayatollah, Khomeini's maverick politics, his charismatic persona, and a series of circumstances would catapult him into the leadership of the revolution and, subsequently, of the Islamic Republic of Iran. While the majority of Iran's *ulama* had remained apolitical, Ruhollah Khomeini had become a strident voice for reform. He had vigorously spoken out against the Shah's policies during 1963–65, been arrested and imprisoned several times, and then exiled to Turkey, to Iraq, and finally at the very end to France. Throughout his exile Khomeini continued to criticize the Shah. Copies of his writings and speeches as well as tape cassettes were smuggled into Iran and circulated through the mosque network.

Khomeini shared with other Islamic ideologues, such as Mawlana Mawdudi of the Jamaat-i-Islami and Hassan al-Banna of the Muslim Brotherhood, a comprehensive vision of Islam:

> Islam has a system and program for all the different affairs of society: the form of government and administration, the regulation of peoples' dealings with each other, the relations of state and people, relations with foreign states and all of the political and economic matters. . . . The mosque has always been a center of leadership and command, of examination and analysis of social problems.[43]

He also shared a polarized view of the world, a world torn between East and West, and an Islamic world facing a crusading West. For Khomeini, the world was divided into two groups: oppressors (the United States and the West in general, as well as the Soviet Union) and oppressed (Muslims and the the Third World).[44] The majority of governments in the Muslim world and the Third World in general were seen as client states, lackeys of the West and the East, hence the revolution's slogan, "Neither East nor West, only Islam." Khomeini's view of the world was particularly colored by his hatred of Western colonialism and imperialism: "the foul claws of imperialism have clutched at the heart of the lands of the people of the Quran, with our national wealth and resources being devoured by imperialism . . . with the poisonous culture of imperialism penetrating to the depths of towns and villages throughout the Muslim world, displacing the culture of the Quran." Iran's relations with Israel, which he regarded as an American colonial outpost, came under blistering attack:

> The sinister influence of imperialism is especially evident in Iran. Israel, the universally recognized enemy of Islam and the Muslims, at war with the Muslim peoples for years, has, with the assistance of the despicable government of Iran, penetrated all the economic, military, and political affairs of the country; it must be said that Iran has become a military base for Israel which means, by extension, for America.[45]

By the seventies Khomeini's original demand for reform was transformed into a revolutionary call for jihad. He denounced Iran's monarchy as illegitimate and anti-Islamic and called for a government guided, if not ruled, by the clergy. In a series of lectures collectively entitled *Islamic Government* which did not become widely known at the time, Khomeini rejected monarchy as un-Islamic, maintained that the Islamic character of a government is determined by Sharia rule, and called for its reimplementation in place of the man-made foreign codes which Iran had adopted. Given the centrality of Islamic law, Khomeini argued, the *ulama* played a necessary advisory and supervisory role for temporal rulers. He even hinted at the possibility of direct clerical rule.

Khomeini's views on Islamic government were not widely known or understood and would have remained academic, had the political situation in Iran not deteriorated so rapidly. The Shah's repressive policies, crushing any and all forms of dissent and leading to the indiscriminate arrest, imprisonment, torture, and death of many at the hands of the secret police, created martyrs, set in motion a cycle of regime and antiregime violence, and provoked the rapid growth of a broadly based resistance movement. In the midst of a widespread suppression of intellectuals, journalists, politicians, liberal nationalists, socialists, and Marxists, Khomeini, with his vitriolic condemnation of the Shah and call for a new political order which would include a constitutional government and socioeconomic reform, became the symbol and center of opposition.

Iran's "Islamic" Revolution

As opposition mounted within Iran, Iran's Shii heritage offered a common set of symbols, a historic identity, and a value system—an indigenous, non-Western alternative, an ideological framework within which a variety of factions could function. The mullah–mosque network served as the leadership and organizational backbone for opposition to the government. Iran's thousands of mosques, scattered throughout every city and village, provided a natural, informal nationwide communications network. As in the Tobacco Protest and the Constitutional Revolution, mosques served as centers for dissent, political organization, agitation, and sanctuary. The government could ban and limit political meetings and gatherings, but it could not close the mosques or ban prayer. The clergy and their students represented a vast reservoir of grass-roots leadership; at the weekly Friday communal prayer, mosque and sermon were transformed into a religiopolitical event and platform, attracting thousands and often resulting in political demonstrations as the faithful left the mosques to return home. The influence of the clergy was joined to that of Islamic lay reformers like Shariati and Bazargan, who enjoyed the respect of many, especially of an increasingly alienated and militant younger generation.

Historically, Shii Islam had been both revolutionary and quietist. In the early centuries of Islamic history, Shii movements had threatened the stability of the caliphate and been a catalyst in the fall of the Umayyads. However, much of Shiism had settled later into a more apolitical quietism reinforced by a messianic belief in the eventual return of the Twelfth Imam. Thus Shii Islam, like all religious traditions, was capable of multiple interpretations. In Iran in the seventies, Shiism moved from quietism to revolutionary activism.

The defeat of Shii forces and the martyrdom of Husain at Karbala by the army of the caliph Yazid in 680 C.E., the paradigmatic event in Shii history, took on special significance, providing the inspirational model for the Iranian Revolution. Husain's martyrdom symbolized the role of Shii Islam as a protest movement in which a small righteous party struggled against the overwhelming forces of evil; it symbolized the battle between good and evil, the forces of God and of Satan, the oppressed and the oppressor, and the revolt of the disinherited. For many Shii, the Shah and his overwhelming military army, like Yazid's army, represented the evils of corruption and social injustice. Like Husain and his forces, the righteous had a religious right and duty to revolt against this modern-day Yazid and undertake a holy war to restore God's law and social justice. Self-sacrifice and even death in God's path (for God and country!) were to be freely accepted, for to die in God's struggle was to become a martyr and win eternal reward.

One Iran or Many Irans?: Diversity and Difference

Heterogeneous groups in the political spectrum, from secularists to Islamic activists, from liberal democrats to Marxists, joined together under the umbrella of Islam. However, beneath the seeming unity of purpose—opposition to the Shah and the desire for a more indigenously rooted modernity—lay a variety of religious and political outlooks and competing agendas. The political opposition ranged from those for whom an Irano-Islamic alternative simply meant Iran's cultural heritage and values, to those who wished to see the establishment of an Islamic state and society. Similar differences existed among the Islamically committed. Clergy as well as laity differed sharply in their Islamic ideologies, their views of the nature of the new political order and its leadership. Nowhere was this clearer than in the juxtaposition of the Ayatollah Khomeini and Dr. Ali Shariati—Khomeini who embodied clerical authority and power; Shariati and other Islamic modernists who represented a far more nonclerical, innovative, creative reformist approach. These differences would emerge in the aftermath of the revolution.

As public political protests and demonstrations mounted in 1978, so did the repressive measures of the state. Events came to a head in Teheran on Black Friday, September 8. Unable to break up a demonstration, the military and police fired upon the crowd of seventy-five thousand from helicopter gunships and tanks which, many would not forget, were purchased from the West and manned by the Western-trained Iranian military. Black Friday was a turning point in the revolution: white- and blue-collar workers, traditional and modern middle classes, city dwellers and rural peasants swelled the ranks of the opposition and engaged in political action. Women who had worn modern attire now joined their more traditional sisters in donning the veil as a symbol of protest against a monarch whose modernization program had once attempted to ban it.

In December, during the sacred month of Muharram which commemorates Husain's martyrdom, Islamic symbol and ritual were fused with contemporary political realities as religious processions became protest demonstrations. In Teheran almost two million people called for the overthrow and death of the Shah, the creation of an Islamic government, and the return and leadership of Khomeini. On January 6, 1979, the Shah, buffeted by widespread dissent and violence, unable to count on a military whose soldiers were defecting, and finding his American patrons wavering in their support, left Iran.

From Monarchy to Islamic Republic

The euphoria of victory and the sense of revolutionary solidarity proved fleeting. Few Iranians were prepared for what unfolded during the early years of Iran's Islamic Republic. These were revolutionary times. For historical and cultural reasons, Islam had emerged as the most effective means for political mobilization. With the fall of the Shah, deep differences in worldviews and interests quickly came to the fore. Few were prepared for the power struggle that would ensue. Those who expected that the mullahs would quietly go back to their mosques or settle into an advisory role to the new government were in for a rude awakening.

Khomeini's criticisms and denunciation of the Shah and his call for a new and just political and social order had been framed in religiocultural categories not unlike those of many other critics of the Shah, and thus masked the deep differences in outlook and agenda which existed. Furthermore, his advisers in Paris and in the early days after his return to Teheran included modern laymen influenced by the outlook of Ali Shariati. Although the opposition had a common enemy (the Shah, Pahlavi despotism, foreign control) and a common purpose (a more just and egalitarian government), there had been no agreement upon the particular form of government or even its leadership.

Few were familiar with Khomeini's writings on the nature of Islamic government. In particular, his belief in direct clerical rule, guardianship or government by a jurist or expert in Islamic law (*vilayat-i-faqih*), was neither a prominent Shii doctrine nor one which had enjoyed widespread support among Iran's religious establishment. Thus few foresaw the control which he and the *ulama* would exert in their creation of a clergy-dominated Islamic government. The implications of this doctrine were only to be realized during the post-1979 period, when Khomeini asserted his role as overseer of Iran's government.

From its inception in February 1979, the provisional government of the new Islamic Republic of Iran witnessed a struggle between moderates and militants. The authoritarianism of the Shah and his modern secular elites was replaced by that of the Imam as Guardian and his clerical cohorts. The coalition which had brought about the revolution disintegrated under the swift and often harsh judgments of a new authoritarianism. Not only Pahlavi officials and supporters but all who differed with the new "clerical autocracy"—secular and religiously oriented, lay and clerical, women and religious minorities—felt the swift arm of Islamic justice, meted out by revolutionary guards and courts. Prisons were again filled, and summary trials and executions got so out of hand that the Ayatollah Khomeini himself intervened and warned against such excesses. The Islamic Republic's first prime minister, Mehdi Bazargan, resigned in disgust; Bani-Sadr, its first elected president, fled to exile in France; Sadeq Gobtzadeh, who had held a number of government posts, was executed for his participation in an alleged plot to assassinate Khomeini. All had been protégés of Khomeini. Those religious leaders who did not accept Khomeini's doctrine of "rule by the jurist" were hounded and harassed by fellow clerics. The Ayatollah Shariatmadari, a senior ayatollah revered for his learning and piety, was himself "defrocked" in the spring of 1982.

The *ulama* and/or their supporters consolidated their control of the government, parliament, the judiciary, the media, and education. Iran's revolution was successfully institutionalized. Government purges (of political organizations, the military, the judiciary, educational institutions, and government bureaus), control of the media, and intimidation of dissident clergy severely restricted the opposition. Similarly, the cultural revolution was promoted and institutionalized in the schools and the media, while alternative viewpoints were restricted or suppressed. Just as the Shah had tried to mandate Western dress, so Islamic dress codes were now enforced on the streets, in government offices, and in the universities.

Clerical and lay voices of dissent were offset by a critical mass of committed clergy and laity. The militant clergy were supported by the "lay stratum of the Islamic republic," who accepted clerical rule, served in the government, and filled many key positions in the bureaucracy.[46] Most shared a common social status (lower middle class and *bazaari* families) and educational background (first-generation university graduates in the sciences and medicine).[47] Thus the Islamic republic was based upon a clerical–lay alliance committed to the Imam and the revolution, though with differences of vision and policy.

The Islamic Threat: Iran's Export of the Revolution

Institutionalization of the revolution was accompanied by a twin goal, export of Islamic revolution. Fear of its export dominated much of Middle Eastern politics for more than a decade.[48] It had a significant impact on both the Muslim world and the West:

> For some, it has been a source of inspiration and motivation; for others, revolutionary Iran has symbolized an ominous threat to the stability of the Middle East and the security of the West because it has been associated with terrorism, hostage-taking, attacks on embassies, and the promotion of revolutionary activities. Indeed, for the Reagan administration, Iran often seemed synonymous with worldwide terrorism and revolution.[49]

Amid the hysteria of the postrevolutionary period, assessing the Iranian threat, separating fact from fiction, proved difficult if not impossible for the West and its allies. The shock of a revolution which had made the unthinkable a reality resulted in an overcompensation that saw both Iranian domestic politics and foreign policy through the lenses of Islamic radicalism and extremism. Intolerant Islamization at home was complemented by the threat of the spread of worldwide terrorism. Fear of other Irans and of a fundamentalist attempt to spread revolution not only in the Muslim world but also throughout the Third World were fed by the public rhetoric and propaganda of Khomeini's Islamic Republic, the outrage of an "America Held Hostage" (the seizure and captivity of the American embassy and its staff), Shii disturbances in the Gulf, and warnings from both Western and Muslim governments of the dangers of radical fundamentalism. The initial euphoria among many in the Muslim world, especially other Islamic movements, merely solidified the image of an expansionist Islamic revolution poised to menace the world. At times there seemed little exaggeration in saying that—for differing reasons—when Khomeini spoke, the whole world listened.

Export of a revolutionary Islam sprang from the Ayatollah Khomeini's ideological worldview, an interpretation of Islam which combined a religiously rooted brand of Iranian nationalism with a belief in the transnational character and global mission of Muslims to spread Islam through preaching and example as well as armed revolution. The promotion and spread of Islam was a primary foreign-policy objective, reflected in the exhortation, in the constitution of the Islamic Republic of Iran, "to perpetuate the revolution both at home and abroad."[50] Khomeini advocated peaceful means like preaching and propaganda, as well as confrontation and armed struggle: "We want Islam to spread everywhere . . . but this does not mean that we intend to export it by the bayonet. . . . If governments submit and behave in accordance with Islamic tenets, support them; if not, fight them without fear of anyone."[51]

The Ministry of Religious Guidance was charged with providing preachers and publications, conducting conferences for *ulama* from overseas, and distributing propaganda abroad. At the same time the Ayatollah Khomeini and broadcasts of Iran's Voice of the Islamic Revolution called upon Muslims of the Gulf and throughout the world to rise up against their governments. He denounced the monarchies of the Gulf and many other Muslim governments both because monarchy was unIslamic and because they had ties with the United States, an influence which he sarcastically dismissed as "American Islam."

Iran's revolutionary impact on other Muslim countries varied significantly. In two countries—Lebanon and Bahrain—its intervention was

direct, tangible, and significant. In many others its influence was more indirect, inspirational, and motivational, encouraging and confirming preexisting Islamic political tendencies and stimulating political thinking from Egypt to the southern Philippines. Though more often than not indirect, Iranian influence was often exaggerated both by disproportionate media coverage and by the tendency of Muslim governments (Egypt, the Sudan, Indonesia, Tunisia, Iraq) to use the Iranian threat as a pretext to discredit or suppress domestic Islamic opposition and as a means to encourage aid from Western governments.

Divisions Within the Government

Criticism of the Shah's Western-inspired modernization program and calls for redressing the oppression of the poor had provided a common ground for opposition, but postrevolutionary attempts to implement an Islamic order revealed substantial ideological and policy differences. Islam was used to justify both state control of the economy and private-sector freedom. While all accepted the authority of Islamic law, competing "Islamic" policies resulted from differences in religiolegal interpretation and conflicting class interests. Some insisted that answers must be found in or based upon explicit texts in traditional Islamic jurisprudence, i.e., past legal interpretations or regulations. Others argued that new problems required new interpretations of God's revelation. Confusion and indecision characterized much of the attempt to institute substantive social reform. A majority in parliament attempted to implement a social revolution. Laws were passed to control prices and markets; to nationalize many industries, banks, and foreign trade; to expropriate urban land for use by the poor and homeless; and to undertake a major redistribution of agricultural lands. However merchants, who had been a major source of financial support for the revolution, and landowners including clerical leaders strongly lobbied politicians and senior clerics against such measures. The Council of Guardians (a committee of clerical experts in Islamic law who determined whether or not parliamentary laws were Islamically acceptable) vetoed much of the reform legislation.

Sharp political differences arose in 1988 in the aftermath of the Iran–Iraq war. That war, while devastating in its human and economic tolls, had provided an excuse for Iran's serious economic problems, minimized criticism of the government, and mobilized support for the regime. Disillusion with the failure to achieve victory, exhaustion of the economy by the war, a general deterioration in the quality of life, and growing public disaffection—all were exacerbated by ideological differences within the government over strategies for national reconstruction and export of the revolution.

The pragmatist camp espoused a policy of national reconstruction and normalization of relations with the West, though the United States remained a special problem. Iran's export of the revolution was to be through example only, rather than political intervention. Radicals or revolutionary purists repudiated any talk of normalization with the West and any retreat from international revolutionary political activism as once again opening the door to dependency on the West.

The tide abruptly turned in February 1989 with the Ayatollah Khomeini's call for the execution of Salman Rushdie, author of *The Satanic Verses*. The Rushdie affair provided the radical camp with an issue to counter the ascendancy of the pragmatists; enabled Khomeini to reassert his Islamic leadership internationally, mobilizing militant fervor for the defense of Islam; and distracted attention from Iran's pressing socioeconomic problems and growing social discontent.

Despite their differences, the death of the Ayatollah Khomeini in June 1989 did not precipitate a catastrophic power struggle between pragmatists and radicals. They closed ranks to assure a smooth transition that would foil the hopes of their enemies. Hashemi Rafsanjani and the Ayatollah Khamenei emerged as the key players. Rafsanjani was elected the new president of Iran, and the Assembly of Experts elected former President Khamenei Iran's "guardian" jurist. Iran's new leadership steered a more flexible course, working together for economic reconstruction and normalization of diplomatic and economic ties with the international community. Rafsanjani cut back on Iran's support for more radical Shii factions in Lebanon, sought the release of American hostages in Lebanon, and pursued diplomatic and economic relations with the West.

For almost two decades Islam has reemerged as a source of government legitimacy and national development. The pervasive belief that nation building required a clear secular path and orientation has been challenged across the Muslim world. As we have seen, governments have appealed to Islam to enhance their authority, buttress nationalism, legitimate policies and programs, and increase popular support. However, a review of contemporary Islamic politics at the state level challenges perceptions of a monolithic Islamic threat. State implementation of Islam has varied markedly in terms of its forms of government, domestic programs, and foreign policies. Monarchs, military rulers, presidents, and clergy have ruled governments as diverse as Saudi Arabia's conservative monarchy, Libya's populist socialist state, Iran's clerical republic, and the Sudan and Pakistan's military regimes. Within specific countries, contending voices and groups have vied for power in the name of Islam. Moreover, the appeal to Islam has also served as a double-edged sword. Those who wield it run the risk of being judged by that very Islamic yardstick, challenged or toppled. While state use or manipulation of Islam can broaden a ruler's base of support, it also draws Islamic as well as secular critics. Sadat and Qaddafi's appeals to Islam increasingly set them on a collision path with Islamic organizations.

Internationally, the Islamic state has not necessarily led to the strengthening of bonds of unity or to a Pan-Islamic threat. Sadat, the "believer-president," had no qualms about supporting the Shah and denouncing the Ayatollah Khomeini as a madman. Egypt's turn to Islam did not prevent the Organization of the Islamic Conference or the Arab League and most Arab and Muslim governments from breaking diplomatic relations after Egypt signed the Camp David Accords. Qaddafi's Islamic period paralleled his cool relations both with Sadat's Egypt and Nimeiri's Sudan. A lesson to be learned is that while Islam can influence a Muslim nation's attitude toward greater cooperation or foreign aid, in most cases the overriding determinant will be national interest. This also accounts for the diversity of relationships that have existed between Islamically oriented governments and the West. Sadat's Egypt, Nimeiri's Sudan, Zia ul-Haq's Pakistan, and King Fahd's Saudi Arabia all enjoyed close ties with the United States, while Qaddafi's Libya and Iran have more often than not been among its severest critics. The extent to which specific leaders and circumstances rather than Islam affect policy can be seen in the Sudan and Pakistan. In the Sudan, Islamic governments have been introduced by Gaafar Nimeiri and more recently by Omar al-Bashir. The former enjoyed close relations with the United States, while under the latter relations have deteriorated markedly.

The imposition of Islam from above has raised many questions: Whose Islam? What Islam? Why a negative Islam? In answer to the first question, Islam—like all ideologies, secular and religious alike—can be a positive social force. It can also be manipulated by dictators and demagogues and, as will also be seen in the next chapter, by Islamic movements. Is the nature and implementation of Islam simply to be left to rulers or to a parliamentary process? More often than not, rulers and governments have used Islam, as they have used other ideologies, to solidify power rather than to promote political participation. Heads of state (kings, generals, former military officers) have found it a useful tool. At the same time many of the clergy, who see themselves as the protectors of Islam, and contemporary Islamic organizations have sought their turn in power. However, increasingly in the Muslim world there is a demand for greater political participation, for governance not simply by strong rulers but through elected parliamentary bodies.

To respond to the second question, "What Islam?," the call for more Islamically oriented societies and the experiments undertaken thus far also raise questions about the nature and direction of Islamic renewal and reform. Is the implementation of Islam in state and society to be a restoration or a reformation, the resurrection of past doctrines and laws or the reconstruction of new models rooted in faith but appropriate to the changed circumstances of life today? The issue is clear when we look at the question of Islamic law. For many, the Islamic character of the state is determined by the implementation of Islamic law. However, differences of interpretation abound. For some Muslims, this simply means the wholesale reapplication of classical or medieval laws. Others argue that, just as the laws delineated in the early Islamic centuries were the product of revelation and human interpretation, conditioned by historical and sociopolitical contexts, so too today Muslims must distinguish between that which is immutable and a vast body of legal regulations that are subject to reinterpretation and reform. The battle between traditionalists and modern reformers is one which all faiths have witnessed, made more acute when its results are to be applied not only to personal but also to public life.

Finally, the significance of the questions "Whose Islam?" and "What Islam?" are highlighted by that of "Why a negative Islam?" For many observers within and without the Muslim world, the implementation of Islam politically has too often seemed less a process of inspiration, renewal, and unity than a means for political and social manipulation and control. Restrictive laws, summary courts, and Islamic punishments and taxes, rather than social and political reform and greater emphasis on human rights, have prevailed. Both rulers who would implement Islam from above, and populist movements that wish to do so from below, are increasingly challenged by those in their societies who ask how an Islamically oriented state and society will be more effective than secular regimes in issues of political liberalization and participation, socioeconomic reform and cultural authenticity. Or, to put it bluntly, is the choice simply between secular and Islamic autocracy?

5

Islamic Organizations:
Soldiers of God

Modern Islamic organizations have been the driving force behind the dynamic spread of the Islamic resurgence. They have also been the focal point or embodiment of the Islamic threat in the eyes of Western governments as well as many governments in the Muslim world. For some, Islamic movements represent an authentic alternative to corrupt, exhausted, and ineffectual regimes. For many others, they are a destabilizing force—demagogues who will employ any tactic to gain power. The violence and terrorism perpetrated by groups with names like the Party of God, Holy War, Army of God, and Salvation from Hell conjure up images of religious fanatics with a thirst for vengeance and a penchant for violence who will stop at little. The assassination of Anwar Sadat, the taking of hostages in Lebanon, and the hijacking of planes embody a "Sacred Rage" that has become all too familiar.[1] Yet again the reality is far more complex than its popular image. The majority of Islamic organizations would claim that, where permitted, they work within the political system and seek change from below through a gradual process of reform. Many Islamic organizations today espouse political liberalization and democratization. Their members participate in elections and serve in legislatures and cabinets.

Given the pivotal role that Islamic movements have played and continue to play both in the politics of the Muslim world and in the plans, calculations, and responses of Western governments, understanding their nature, goals, and activities becomes critical in assessing Islamic movements and the Islamic threat.

While Islamic revivalism and Islamic movements are integral to Islamic history and in some sense may be seen as part of a recurrent

119

revivalist cycle in history, most movements today differ from those of earlier centuries in that they are modern, not traditional, in their leadership, ideology, and organization. If we speak of fundamentalism as a return to the foundations of Islam, the Quran, and the example of the Prophet in order to renew the community, then these movements are neofundamentalist or neorevivalist, for they look to the sources of Islam not simply to replicate the past but to respond to a new age.

The Muslim Brotherhood and the Jamaat-i-Islami

In trying to understand the origins and nature of modern Islamic movements, two organizations in particular dominate the landscape of the Muslim world in the twentieth century. Contemporary Islamic activism is particularly indebted to the ideology and organizational example of the Muslim Brotherhood and the Jamaat-i-Islami (Islamic Society or Group). Their founders and ideologues, Hassan al-Banna and Sayyid Qutb of the Brotherhood and Mawlana Abul Ala Mawdudi of the Jamaat, have had an incalculable impact on the development of Islamic movements throughout the Muslim world. They are indeed the trailblazers or architects of contemporary Islamic revivalism, men whose ideas and methods have been studied and emulated from the Sudan to Indonesia.

Hassan al-Banna (1906–49), a schoolteacher and former disciple of the Islamic modernist Rashid Rida, established the Muslim Brotherhood (Ikhwan al-Muslimin) in Egypt in 1928, while Mawlana Abul Ala Mawdudi (1903–79), a journalist, organized the Jamaat-i-Islami in India in 1941.[2] Both movements arose and initially grew in the thirties and forties at a time when their communities were in crisis. Both blamed European imperialism and a Westernized Muslim leadership for many of the current problems.

In Egypt the Brotherhood's critique of Western imperialism and the ills of Egyptian society in time found a receptive audience among the religiously inclined as well as the more Western, secular-oriented elites. Initial faith in liberal nationalism had been shaken by the defeat of the Arabs in Palestine, the creation of the state of Israel with British and American support, Egypt's continued inability to shake off British occupation, and massive unemployment, poverty, and corruption. Muslim Brothers greatly enhanced their credentials as patriotic sons of Egypt and Arab nationalists in their significant participation in the 1948 Palestine war and again in the 1951 Suez crisis.

In South Asia Mawlana Mawdudi had witnessed the collapse of the Ottoman empire and the failure of the Caliphate Movement to save that empire from being dismembered by Britain and France for its support of Germany during World War I. Growing Hindu assertiveness in the Indian Freedom (Independence) Movement in South Asia contributed to Mawdudi's perception of the continued deterioration of Muslim power and the threat to Islam and the Muslim community. Mawdudi blamed European colonialism and the emergence of modern nationalism, a foreign and Western ideology which divided rather than united peoples, replacing the universal or Pan-Islamic ideal and solidarity with a more tenuous and divisive identity based upon language, tribe, or ethnicity.

Hassan al-Banna and Mawlana Mawdudi, who were contemporaries, were pious, educated men with traditional Islamic backgrounds and knowledge of modern Western thought. Both saw their societies as being dependent on the West, politically weak, and culturally adrift. Each in his early years had been an anticolonial nationalist who turned to religious revivalism to restore the Muslim community at home and universally. These founders and ideologues of their organizations drew on the example and concerns both of eighteenth-century Islamic revivalist movements like the Wahhabi of Saudi Arabia and nineteenth- and twentieth-century Islamic modernist predecessors for their critique of Muslim society, their revivalist and reformist worldview emphasis on organization, and their sociopolitical activism. They did not simply retreat to the past but instead provided Islamic responses, ideological and organizational, to modern society.[3] Hassan al-Banna and Mawlana Mawdudi appropriated and reapplied the vision and logic of the revivalist tradition in Islam to the sociohistorical conditions of twentieth-century Muslim society. In a very real sense they modernized Islam by providing a modern interpretation or reformulation of Islam to revitalize the community religiously and sociopolitically. However, they did this not by purifying Islam of cultural accretions or un-Islamic beliefs and values or by restoring early practices of the community, however authentic. They self-consciously reapplied Islamic sources and beliefs, reinterpreting them to address modern realities. Yet they distinguished their method from that of Islamic modernism, which they equated with the Westernization of Islam. If Islamic modernists legitimated the adoption of Western ideas and institutions by maintaining their compatibility with Islam, al-Banna and Mawdudi sought to produce a new synthesis which began with Islamic sources and found either Islamic equivalents or Islamic sources for notions of government accountability, legal change, popular participation, and educational reform.

Hassan al-Banna and Mawlana Mawdudi shared a common anti-imperialist view of the West, which they believed was not only a political and economic but also a cultural threat to Muslim societies. Westernization threatened the very identity, independence, and way of life of Muslims. Indeed, they regarded the religiocultural penetration of the West (education, law, customs, values) as far more pernicious in the long run than political intervention, since it threatened the very identity and survival of the Muslim community. Imitation of or dependence on the West was to be avoided at all cost. The Brotherhood and the Jamaat proclaimed Islam as a self-sufficient, all-encompassing way of life, an ideological alternative to Western capitalism and Marxism. They joined thought to action, creating organizations that engaged in political and social activism.

Though hostile to Westernization, they were not against moderniza-tion. Both Hassan al-Banna and Mawlana Mawdudi engaged in modern organization and institution building, provided educational and social welfare services, and used modern technology and mass communications to spread their message and to mobilize popular support. Their message itself, though rooted in Islamic revelation and sources, was clearly written for a twentieth-century audience. It addressed the problems of moder-nity, analyzing the relationship of Islam to nationalism, democracy, capi-talism, Marxism, modern banking, education, law, women and work, Zionism, and international relations. Mawdudi far more than al-Banna wrote extensively and systematically, attempting to demonstrate the com-prehensive relevance of Islam to all aspects of life. The range of his topics reflected his holistic vision: Islam and the state, economics, education, revolution, women.

Despite differences, al-Banna's and Mawdudi's reinterpretations of Islamic history and tradition produced a common ideological worldview which has inspired and guided many modern Islamically oriented so-ciomoral reform movements. This worldview not only governed their organizations but also informed Islamic movements that sprang up throughout the Muslim world in subsequent decades.

Among the primary principles of al-Banna's and Mawdudi's ideologi-cal worldview were the following:

1. Islam constitutes an all-embracing ideology for individual and corpo-rate life, for state and society.

2. The Quran, God's revelation, and the example (Sunnah) of the Prophet Muhammad are the foundations of Muslim life.

3. Islamic law (the Sharia, the "path" of God), based upon the Quran and the Prophet's model behavior, is the sacred blueprint for Muslim life.

4. Faithfulness to the Muslim's vocation to reestablish God's sover-eignty through implementation of God's law will bring success, power, and wealth to the Islamic community (ummah) in this life as well as eternal reward in the next.

5. The weakness and subservience of Muslim societies must be due to the faithlessness of Muslims, who have strayed from God's divinely revealed path and instead followed the secular, materialistic ideologies and values of the West or of the East—capitalism or Marxism.

6. Restoration of Muslim pride, power, and rule (the past glory of Islamic empires and civilization) requires a return to Islam, the reimplementation of God's law and guidance for state and society.

7. Science and technology must be harnessed and used within an Islamically oriented and guided context in order to avoid the Westernization and secularization of Muslim society.

Both the Brotherhood and the Jamaat saw but two choices, darkness or light, Satan or God, ignorance (that which is un-Islamic) or Islam. At the heart of the message of the Brotherhood and the Jamaat was the conviction that Islam provided a divinely revealed and prescribed third alternative to Western capitalism and Soviet Marxism.

Scripture and tradition were appealed to and reinterpreted by the founders of these modern religious societies. The inspiration and continuity of the past were coupled with a response to the demands of modernity. This combination of past and present are demonstrated not only by their reinterpretation of Islam but also by their organization and activities. Organizationally, the Brotherhood and the Jamaat followed the example of the Prophet Muhammad (also emulated by seventeenth- and eighteenth-century revivalist movements) in gathering together believers committed to establishing societies governed by God's law. They were to be a vanguard, a righteous community within the broader community—the dynamic nucleus for true Islamic reformation or revolution, returning society to the straight path of Islam.

Both organizations recruited followers from mosques, schools, and universities: students, workers, merchants, and young professionals. They were primarily urban, based among the lower middle and middle classes, with whom they were especially successful. The goal was to produce a new generation of modern-educated but Islamically oriented leaders prepared to take their place in every sector of society. However, while al-Banna worked to develop a broad-based populist movement, Mawdudi's Jamaat was more of an elite religious organization whose primary goal was to train leaders who would come to power. An Islamic revolution was ultimately necessary to introduce an Islamic state and society. However, this Islamic revolution was to be first and foremost a social rather than a violent political one. The establishment of an Islamic state first required the Islamization of society through a gradual process of social change. Both organizations espoused an "Islamic alternative" to conservative religious leaders and modern, secular, Western-oriented elites. The *ulama* were generally regarded as passé, a religious class whose fossilized Islam and cooption by governments were major causes of the backwardness of the Islamic community. Modernists were seen as having traded away the very soul of Muslim society out of blind admiration for the West.

Though these organizations were quick to denounce the evils of imperialism and the cultural threat of the West, both the Brotherhood and the Jamaat nevertheless realized that the Muslim predicament was first and foremost a Muslim problem, caused by Muslims who had failed to be sufficiently Islamically observant. Its rectification was the primary task of Muslims. Rebuilding the community and redressing the balance of power between Islam and the West must begin with a call or invitation (*dawa*) to all Muslims to return to and reappropriate their faith in its fullness—to be born again in the straight path of God. Both the Brotherhood and the Jamaat reemphasized and interpreted the concept of *dawa*. The call to Islam has two aspects: an invitation to non-Muslims to convert to Islam, and the calling of those who were born Muslim to be better Muslims. Both organizations focused on the latter, the transformation (Islamization) of the individual and society. The reversal of the fortunes of Islam and of the Muslim community was to be accomplished through a social revolution. Religious commitment, modern learning and technology, and activism were combined as the Brotherhood and the Jamaat disseminated their interpretation of Islam through schools, publications, preaching, social services, and student organizations.

Neorevivalism and the West

Like secular and Islamic modernists, al-Banna and Mawdudi acknowledged the weakness of Muslim societies, the need for change, and the value of science and technology. However, they criticized both secular and Islamic modernists for excessive dependence on the West, the cause of the continued impotence of Muslim societies. On the one hand, secularists separated religion from society and took the West as the model for development: "Until recently, writers, intellectuals, scholars and governments glorified the principles of European civilization . . . adopted Western style and manner."[4] Despite their anticolonial politics, secular Muslim leaders were in effect regarded as indigenous Western cultural colonizers. On the other hand, Islamic modernists were also criticized: in their zeal to demonstrate the compatibility of Islam with modernity, they employed or relied upon Western values, producing a Westernized Islam.

All these people in their misinformed and misguided zeal to serve what they hold to be the cause of Islam, are always at great pains to prove that Islam contains within itself the elements of all types of contemporary social and political thought and action. . . . [T]his attitude emerges from an inferiority complex, from the belief that we as Muslims can earn no honour or respect unless we are able to show our religion resembles modern creeds and is in agreement with most of the contemporary ideologies.[5]

In contrast to Islamic modernists, these neorevivalists were more sweeping in their indictment of the West and their assertion of the total self-sufficiency of Islam. They maintained that Muslims should not look to Western capitalism or communism (white or red imperialism) but solely to Islam, the divinely revealed foundation of state and society. The Brotherhood charged that faith in the West was misplaced. Western democracy had not merely failed to check but contributed to authoritarianism (the manipulation of the masses by modern elites), economic exploitation, corruption, and social injustice. Western secularism and materialism undermined religion and morality, society and the family. The inherent fallacy of Western secularism, separation of religion and the state, would be responsible for the West's moral decline and ultimate downfall. Finally, the Brotherhood maintained that, despite Arab subservience to the West, the West had betrayed the Arabs in its support for the Israeli occupation of Palestine: "The Palestine question became the starting-point for attacks on the United States . . . [and resulted in] the full identification of Zionism with crusading Western imperialism."[6] In contrast to Islamic modernists, the goal of the Brotherhood and the Jamaat was not to render Islam compatible with Western culture, but to create a more indigenously rooted, authentic Islamic state and society through a process of renewal or Islamization based upon "a return to the principles of Islam . . . [and] the reconciliation of modern life with these principles, as a prelude to a final Islamization (of society)."[7]

Denunciation of the West did not mean wholesale rejection of modernization. Both al-Banna and Mawdudi distinguished between Westernization and modernization, between Western values and modern ideas and institutions. Thus the best of science and technology as well as political ideals could be appropriated, though selectively and carefully, if separated from Western values contrary to Islam and informed instead by Islamic values. In the final analysis, the renaissance or reformation of Islam would not come from reason and secularism but revelation. As Mawdudi noted:

We aspire for Islamic renaissance on the basis of the Quran. To us the Quranic spirit and Islamic tenets are immutable; but the application of this spirit in the realm of practical life must always vary with the change of conditions and increase of knowledge. . . . Our way is quite different both from the Muslim scholar of recent past and modern Europeanized stock. On the one hand we have to imbibe exactly the Quranic spirit and identify our outlook with the Islamic tenets while, on the other, we have to assess thoroughly the developments in the field of knowledge and changes in conditions of life that have been brought during the last eight hundred years; and third, we have to arrange these ideas and laws of life on genuine Islamic lines so that Islam should once again become a dynamic force; the leader of the world rather than its follower.[8]

Mawdudi's attitude toward democracy provides an excellent example of his method. Since Islam's worldview was God-centered rather than human-centered, a parliamentary democracy based upon popular sovereignty rather than divine sovereignty was unacceptable. Mawdudi rejected democracy—that is, Western democracy, which in the name of majority rule permitted practices such as alcohol consumption and sexual promiscuity that were contrary to God's law. However, parliamentary political participation or consultative assemblies subordinated to Islamic law, God's law, were permissible. Here both al-Banna and Mawdudi reinterpreted and utilized the Islamic concept of consultation (*shura*) to provide Islamic justification. Mawdudi preferred to speak of the Islamic system as a "theodemocracy," as distinct from a theocracy or clerical state, in which the popular will was subordinated to and limited by God's law. Mawdudi did not shrink from those who ridiculed him for "religious authoritarianism." Indeed, Mawdudi did not envision submission to God's absolute authority as a deprivation of man's liberty, but rather as a condition of it. Thus he had no problem characterizing an Islamic government or theodemocracy as "Islamic totalitarianism."

Radical Islam: Sayyid Qutb, Martyr

Just as the ideological worldviews of Hassan al-Banna and Mawlana Mawdudi had been shaped by their social context, so too the ideology of Islamic revivalism in Egypt became more militant and combative in the late fifties and sixties as a result of the Muslim Brotherhood's confrontation with the Egyptian state. One man stands out as the architect of radical Islam, Sayyid Qutb (1906–66). By the sixties Qutb, increasingly radicalized by Nasser's suppression of the Brotherhood, transformed the ideological beliefs of al-Banna and Mawdudi into a rejectionist revolutionary call to arms. Like al-Banna, Qutb would come to be remembered as a martyr of the Islamic revival.

Also like al-Banna, Qutb studied at the Dar al-Ulum, a modern college established by reformers to train teachers in modern subjects. It was here that he became familiar with Western literature and, like many young intellectuals of the time, grew up to be an admirer of the West. After graduation he worked as an official in the Ministry of Public Instruction. He was an active participant in the literary and social debates of his times and soon became a prolific writer, publishing works of poetry and literary criticism. Qutb was also a devout Muslim who had memorized the Quran as a child and in the post–World War I period wrote increasingly on Islam and the state of Egyptian society. In 1948 he published *Social Justice in Islam,* in which he maintained that, in contrast to Christianity and communism, Islam possessed its own distinctive social teachings or path, Islamic socialism, which avoided the pitfalls of separation of religion and society on the one hand and atheism on the other.

> Christianity looks at man only from the standpoint of his spiritual desires and seeks to crush down the human instincts in order to encourage those desires. On the other hand, Communism looks at man only from the standpoint of his material needs; it looks not only at human nature, but also at the world and at life from a purely material point of view. But Islam looks at man as forming a unity whose spiritual desires cannot be separated from his bodily appetites, and whose moral needs cannot be divorced from his material needs. It looks at the world and at life with this all-embracing view which permits of no separation or division. In this fact lies the main divergence between Communism, Christianity, and Islam.[9]

In 1949 Sayyid Qutb traveled to the United States to study educational organization. This experience proved to be a turning point in his life. After this visit he became a severe critic of the West and, shortly after his return to Egypt in 1951, he joined the Muslim Brotherhood. Although he came to the United States out of admiration, Qutb experienced a strong dose of culture shock which drove him to become more religiously observant and convinced of the moral decadence of Western civilization and its anti-Arab bias. He was disgusted by what he judged to be racial prejudice toward Arabs, reflected both in how he and other Arabs were treated and by U.S. government and media support for Israel.[10] He was scandalized by the sexual permissiveness and promiscuity of American society, the free use of alcohol, and free mingling of men and women in public. In his writings

> [h]e describes Americans as being violent by nature and as having little respect for human life. . . . American churches were not places of worship as much as entertainment centers and playgrounds for the sexes. Ameri-

cans, according to Qutb, were primitive in their sexual life, as illustrated in
the words of an American female college student who told him that the
sexual issue was not ethical, but merely biological.[11]

During the fifties Qutb emerged as a major voice of the Muslim Brotherhood and its most influential ideologue. His commitment, intelligence,
militancy, and literary style made him especially effective within the context of a growing confrontation between a repressive regime and the
Brotherhood. Government harassment of the Brotherhood and Qutb's
imprisonment and torture in 1954 for alleged involvement in an attempt
to assassinate Nasser only increased his radicalization and confrontational
worldview. During ten years of imprisonment in the equivalent of a concentration camp, he wrote prolifically, completing *In the Shade of the
Quran,* a Quranic commentary, as well as his most influential Islamic
ideological tract, *Maalim fil Tariq* (*Signposts* or *Milestones*). His thought
now reflected a new revolutionary vision born of his extended imprisonment and torture. He carried the ideas of Hassan al-Banna and especially
Mawlana Mawdudi to their literalist, radical conclusions.

For Qutb, Islamic movements existed in a world of repressive, anti-
Islamic governments and societies. Society was divided into two camps,
the party of God and the party of Satan, those committed to the rule of
God and those opposed. There was no middle ground. Strongly influenced by Mawdudi, Qutb emphasized the development of a vanguard, a
group (*jamaa*) of true Muslims within the broader corrupted and faithless society. The Islamic movement (*haraka*) was a righteous minority
adrift in a sea of ignorance and unbelief (*jahiliyya*). He dismissed Muslim governments and societies as un-Islamic (*jahili*), being in effect atheist or pagan. Thus the classical historical designation of pre-Islamic Arabia as a society of ignorance (*jahiliyya*) was appropriated to condemn
modern societies as un-Islamic or anti-Islamic. For Qutb, the cause was
the displacement of Islam's God-centered universe by a human-centered
world.

Qutb maintained that the creation of an Islamic system of government was a divine commandment, and therefore not just an alternative
but an imperative.[12] Given the political realities of authoritarian, un-
Islamic regimes, Qutb concluded that attempts to bring about change
from within the existing repressive Muslim political systems were futile,
and that jihad was the only way to implement a new Islamic order.
Jihad as armed struggle in the defense of Islam against injustice became the prescribed path for all true believers in the current crisis.
Islam stood on the brink of disaster, threatened by repressive anti-
Islamic governments and the neocolonialism of the West and the East.

Those Muslims who refused to participate or wavered were to be counted among the enemies of God. Qutb's formulation became the starting point for many radical groups.[13] The two options—evolution, a process which emphasizes revolutionary change from below, and revolution, the violent overthrow of established (un-Islamic) systems of government—have remained the twin paths of contemporary Islamic movements.

For Qutb, as for al-Banna and Mawdudi, the West is the historic and pervasive enemy of Islam and Muslim societies, both a political and a religiocultural threat. Its clear and present danger comes not only from its political, military, and economic power, but also from its hold on Muslim elites who govern and guide by alien standards which threaten the identity and soul of their societies. However, Qutb went beyond his predecessors when he declared Muslim elites and governments to be atheists against whom all true believers should wage holy war.

In 1965 the Muslim Brotherhood, blamed for an attempt on Nasser's life, was massively and ruthlessly suppressed by the government. Qutb and several other leaders were arrested and executed, and thousands of Brothers were arrested and tortured, while others went underground or fled the country.

Political Action

The political lives of the Muslim Brotherhood and the Jamaat-i-Islami tell us much about the nature of Islamic movements and their development both in the past and today. In particular, their diverse paths underscore the influence of sociopolitical contexts on the ideologies and politics of movements, their degree of militancy, and their use of violence. Though the Muslim Brotherhood and the Jamaat began as sociomoral reform organizations, they both became heavily involved in politics, convinced that a more Islamic state and society ultimately required the cooperation and support of the state in implementing Islamic law. The Brotherhood of Hassan al-Banna during the rule of King Farouk, and later in the time of Sayyid Qutb under Nasser's regime, were drawn into confrontation with governments that resisted their demands. Both al-Banna and Qutb came to be remembered as martyrs of Islamic revivalism. Hassan al-Banna was killed reportedly by secret police in retaliation for the Brotherhood's reputed involvement in the assassination of an official in King Farouk's government. Similarly, Sayyid Qutb was hanged as the ringleader of an alleged massive Brotherhood conspiracy against the government.

Though ostensibly one organization, the Muslim Brothers had in fact split internally into several factions. After al-Banna's death, no single leader enjoyed his authority. While Qutb's radical rhetoric appealed to a more militant wing, especially disaffected youth, many other Brothers, including their leader or Supreme Guide, were wary of a direct confrontation with the regime. The political strategy of the Brotherhood remained unresolved, with two competing models, evolution and revolution. While some Brothers may have conspired to overthrow the government, many of the older guard, fully aware of the power of the state, preferred to pursue change through preaching and social activism. Since they were divided, they were no real threat to Nasser. However, they did provide a convenient means for Nasser to deflect attention from his domestic and international problems: "The 'new conspiracy of the Muslim Brotherhood' offered an ideal scapegoat that would enable the leader to reunite the people behind him."[14] As we shall see, despite the Brotherhood's suppression in 1954 and its apparent extinction in 1965, it would reemerge in Egypt in the seventies.

Mawlana Mawdudi and the Jamaat have always been embroiled in Pakistani politics; though militant, they have participated within the political system. Although initially objecting to the creation of Pakistan, after independence Mawdudi and the Jamaat became leading voices in the Islamic Republic of Pakistan, functioning as a political party and pressing their demands for an Islamic state. Although Mawdudi and Jamaat leaders confronted the government from time to time and were imprisoned for their activities, on the whole they were prominent in parliamentary politics. The Jamaat has never been able to score impressive victories in electoral politics but has nonetheless had a significant impact, especially during the rule of Gen. Zia ul-Haq (1977–88). Internationally, Jamaat organizations exist and have been politically active in Pakistan, Bangladesh, India, Afghanistan, and Kashmir. Mawdudi's writings have been disseminated throughout the Muslim world and have influenced Islamic activists from Sayyid Qutb to the current generation ranging from Tunisia to Indonesia.

It is difficult to overestimate the impact of al-Banna, Qutb, and Mawdudi. The ideas and strategies of these modern pioneers continue to be major ideological influences on the worldview and development of Islamic organizations today. Combining religiopolitical activism with social protest or reform, contemporary Islamic movements represent a spectrum of positions from moderation and gradualism to radicalism and revolutionary violence, from criticism of the West to rejection of and attack upon all that is associated with it.

Contemporary Islamic Movements in Egypt: Evolution or Revolution?

Within five years after the death of Sayyid Qutb and the suppression of the Muslim Brotherhood, those who had pronounced its demise were stunned to see the phoenix rise from its ashes. Astonishingly, Anwar Sadat, Nasser's vice president and protégé who had sat on military tribunals that condemned Muslim Brothers, and who had succeeded Nasser as president in 1970, became the patron of Egypt's Islamic resurgence, rehabilitating the Brotherhood and fostering Islamic student organizations. However, Islam proved to be a two-edged sword, for Sadat was subsequently assassinated by religious extremists in its name. The fractured and factionalized legacy of the Muslim Brotherhood and the confrontation of its extremist offshoots with the state have set the pattern for Islamic politics for more than a decade. The majority of Muslim organizations, epitomized by the Muslim Brotherhood, pursue an evolutionary path of preaching and sociopolitical action, while a minority of extremists with names like Holy War and Salvation from Hell directly threaten the authority of the state.

The Muslim Brotherhood, 1970–1991

From 1970 to 1991 the Muslim Brotherhood rebuilt its organization, self-consciously espousing a policy of moderate reformism under both Anwar Sadat and his successor, Hosni Mubarak. Initially many dismissed them as a weakened remnant of defeated and cowed old men. This judgment seemed true in the early Sadat years. The government established a working relationship with the Supreme Guide of the Brotherhood, Omar Tilmassani, whom Sadat had freed from prison. Radical groups charged that the Brotherhood's leadership had been broken by their prison experience and coopted by the government. In fact, the moderate faction of the Brotherhood had prevailed while younger members, radicalized by their prison experiences and inspired by a literal and militant interpretation of Sayyid Qutb, formed secret underground groups bent upon the violent overthrow of the regime.

The Brotherhood worked out a modus vivendi with the Sadat government. While still not recognized as a political party, it was once again able to function openly, preaching its message, publishing magazines, establishing social welfare and financial institutions. Muslim Brothers long underground and those in exile now returned to work again with their colleagues who had been released from prison. They continued to draw their members from the middle and lower middle classes: businessmen, bureaucrats, doctors, engineers, lawyers. Revenue poured in from Brothers working in the oil-rich countries of the Gulf and Iraq and from patrons like Saudi Arabia. While the Brotherhood was not silent, during the early Sadat years it attenuated its criticism of the government, emphasizing cooperation. The leadership of the Brotherhood consistently eschewed violence and confrontation and was critical of the Islamic radical groups' use of force. If the period 1945–65 had been marked by sporadic confrontation and antiregime violence, the post-1970 Brotherhood, under its third Supreme Guide, Tilmassani, underwent an unambiguous transformation. It clearly opted for sociopolitical change through a policy of moderation and gradualism which accepted political pluralism and parliamentary democracy, entering into political alliances with secular political parties and organizations as well as acknowledging the rights of Coptic Christians. Even in the late seventies, when it felt the wrath of Sadat for its growing independence (rejection of the Camp David Accords and Muslim Family Reforms, bitter denunciation of the United States and Israel), it nevertheless continued to pursue the path of critic and pressure group. Though sympathetic to many of the concerns of extremist groups, it remained steadfast in its rejection of violence and terrorism and stayed scrupulously within the limits of Egyptian law.

Although the radical groups that challenged and eventually assassinated Sadat received the most attention during the late Sadat period, by the eighties the Brotherhood once again proved to be a significant dynamic force in Egyptian society, possessing both wealth and power. Under Hosni Mubarak's more open political system, Islamic revivalism emerged as part of mainstream Islam. No longer could it simply be dismissed as religious extremism, a radical fringe of alienated, marginalized Egyptians. The revivalist spirit, increased religious consciousness and observance, was evident throughout much of society and had become normalized and institutionalized, as witnessed in personal religious observances, the growth of Sufi mysticism, and the proliferation of Islamic banks and investment houses, social welfare services (schools, day-care centers, hospitals, and clinics), publishing houses, and media. Though many Islamic associations are apolitical, these organizations and activities carry an implicit political critique of the Egyptian government's failure to meet the needs of society. The Egyptian sociologist Saad Eddin Ibrahim has commented:

> This strand of Islamic activism has therefore set about establishing concrete Islamic alternatives to the socio-economic institutions of the state and the

capitalist sector. Islamic social welfare institutions are better run than their state–public counterparts, less bureaucratic and impersonal. . . . They are definitely more grass roots oriented, far less expensive and far less opulent than the institutions created under Sadat's *infitah* (open-door policy), institutions which mushroomed in the 1970's and which have been providing an exclusive service to the top 5% of the country's population. Apolitical Islamic activism has thus developed a substantial socio-economic muscle through which it has managed to baffle the state and other secular forces in Egypt.[15]

The Muslim Brothers reemerged in the eighties during Mubarak's rule as the largest and strongest Islamic organization, relying on both social and political activism. They ran an extensive network of banks, investment companies, schools, medical and legal services, factories, agribusinesses, and mass communications organizations. The Brotherhood was a major force in professional organizations and syndicates (lawyers, engineers, journalists, doctors) and in the universities. Prevented from participating in elections as a political party, the Brotherhood joined with the Wafd party in the 1984 elections and subsequently formed a new coalition, the Islamic Alliance, with the Labor party in the 1987 elections. Campaigning with the slogan "Islam is the solution" and calling for the implementation of Islamic law, they won 17 percent of the vote, emerging as the chief political opposition of Mubarak's government.

Islamic Revolutionary Groups in Egypt

Throughout both the Sadat and Mubarak years, secret Islamic revolutionary groups waged jihad to disrupt society and challenge the authority and legitimacy of the state. Weapons, bombs, armed confrontation, kidnapping, and assassination have been an integral part of their action. Among the major groups were the Islamic Liberation Organization (also known as Shabab Muhammad, or Muhammad's Youth), Jamaat al-Muslimin (Society of Muslims) or, as it was more popularly known, Takfir wal-Hijra (Excommunication and Emigration); Jamaat al-Jihad (Holy War Society); and Salvation from Hell.

The Islamic Liberation Organization and Takfir wal-Hijra sprang up after the 1967 Arab–Israeli war. The humiliating Egyptian defeat and the loss of Jerusalem were taken as clear signs of a politically impotent, inept, and corrupt system of government. While the Muslim Brothers moderated their voices during the early Sadat years, this new generation of Islamic militants, some of whom had been among the younger members of the Brotherhood imprisoned and tortured during the Nasser

years, espoused a violent antigovernment strategy. The Islamic Liberation Organization attempted a coup d'état in April 1974. Although it
seized Cairo's Technical Military Academy, government forces foiled its
attempt to assassinate Anwar Sadat and declare an Islamic republic. In
July 1977 Takfir wal-Hijra kidnapped Husayn al-Dhahabi, a teacher at
al-Azhar University and former minister of religious endowments, who
had been a strong critic of extremists. They subsequently killed him
when their demands were not met. In retaliation, the leaders of both
Muhammad's Youth and the Takfir were executed by the government,
and many of their members were tried before military courts and imprisoned.[16] Although suppressed, many militants again went underground
and became active in other radical groups such as the Jund Allah (Army
of God) and Jamaat al-Jihad (Holy War Society).

Jamaat al-Jihad, which assassinated Anwar Sadat on October 6, 1981,
developed from the survivors of the abortive coup staged by Muhammad's Youth in 1974. Separate groups grew up around local leaders in
Upper Egypt (Asyut, Minya, Fayyum) as well as in Cairo and Giza.[17]
Members were drawn from the presidential guard, military intelligence,
civil servants, radio and television workers, and university students and
professors. In 1980 the different groups were brought together in a
loosely organized movement governed by a consultative council. All
were united in their belief that the establishment of an Islamic society
required the restoration of the caliphate. All Muslim rulers were regarded as apostates: "The Rulers of this age are in apostasy from Islam.
They were raised at the tables of imperialism, be it Crusaderism, or
Communism, or Zionism."[18] Thus holy war against Egypt's "atheist"
ruler and state was considered both necessary and justified.

Jihad: The Forgotten Obligation

The group al-Jihad's ideology was set forth in a brief tract by Muhammad al-Farag. *The Neglected Obligation* (*Absent Commandment*, i.e.,
jihad) articulated the obligation of jihad, drawing heavily on the ideological worldview of al-Banna, Mawdudi, and Qutb. Farag took their
thoughts on Islamic revolution literally and, following Qutb, pushed
them to their logical conclusion. At the heart of al-Jihad's message and
mission was the call to true believers to wage holy war against Egypt's
un-Islamic state and its leader, Anwar Sadat. It maintained that jihad
was the sixth pillar of Islam, a fact often forgotten or obscured by the
ulama and most Muslims.

Jihad . . . for God's cause, in spite of its extreme importance for the future of religion, has been neglected by the *ulama* . . . of this age. They have feigned ignorance of it, but they know that it is the only way to the return and the establishment of the glory of Islam anew. . . . There is no doubt that the idols of this world can only disappear through the power of the sword.[19]

The goal was the creation of an Islamic state, the eradication of Western law, and the imposition of Islamic law. Al-Jihad believed that radical surgery was required; power must be seized. "We have to establish the Rule of God's Religion in our own country first, and to make the Word of God supreme. . . . There is no doubt that the first battlefield for jihad is the extermination of these infidel leaders and to replace them by a complete Islamic Order. From here we should start."[20]

Despite the imprisonment and execution of al-Jihad leaders for the murder of Sadat, a small group called Salvation from Hell, an offshoot of al-Jihad, resurfaced. Salvation from Hell rejected anyone associated with the "atheist" regime of Mubarak. They declared a holy war against Egypt's "infidel" rulers and society and sought to create an Islamic state.[21] In September 1989 members of Salvation from Hell were sentenced for the attempted assassination of two former cabinet ministers and a journalist who had written articles condemning religious extremists.

The Ideological Worldview of Islamic Militants

The radical confrontational worldview of Islamic militants incorporated the polarized anti-Western perception of the world preached by Sayyid Qutb in the last years of his life. The Westernization of Muslim society was blamed for political corruption, economic decline, social injustice, and spiritual malaise in Egyptian society. Following Qutb, they likened the condition of Egyptian society to the ignorance, paganism, and barbarism prior to Islam. The West's Crusader mentality and neocolonialism and the power of Zionism were believed to be behind a Judaeo-Christian conspiracy which pitted the West against the Islamic world.

Since the legitimacy of Muslim governments is based on Islamic law, militants believed that Egypt's failure to implement the law rendered their country an "atheist state" against which all true Muslims were duty-bound to wage jihad. Jihad against all unbelievers is a religious duty; militants reinterpreted Islamic beliefs, maintaining that true believers are obliged to fight those Muslims who do not share their total commitment and that non-Muslim "People of the Book" were also to be regarded as infidels.

Militants were equally harsh in their denunciation of the religious establishment and their government-supported and regulated mosques. The official *ulama* were regarded as puppets of the government. Their quiescent interpretation of jihad, which downplayed armed struggle and limited jihad to the pursuit of virtue, compromised the true revolutionary meaning of Islam by preaching subservience to the state. Moreover, the officials of al-Azhar could be counted on to support government policies from Camp David and the denunciation of the Ayatollah Khomeini to the condemnation of militant Egyptian groups. Thus Shaykh Abdul Halim Mahmud, rector of al-Azhar, had declared that young activists had been "seized by the devil," and that "the religious and political authorities had to defend the world of Islam against such heresies."[22] Militants told their members to shun state-supported and controlled mosques as places of unbelief, since God's will and the Prophet's teachings were not upheld there. The Muslim Brotherhood was also rejected. Their moderate tone and agenda, their advocacy of a gradual Islamization of Egyptian society, were seen as unrealistic and as a capitulation to the government. Compromise was regarded as collaboration with the enemies of God.

Leadership and Organization

The leadership and organization of Egypt's radical Islamic organizations have reflected the temperament and styles of their leaders. Shukri Mustafa ran the Takfir as a highly disciplined organization, controlled and guided by its Amir. Like Muhammad's Youth, Takfir viewed contemporary Egyptian society as un-Islamic, a domain of unbelievers. Emulating the Prophet Muhammad who, when faced with the unbelief of Mecca, had emigrated to Medina, Takfir set up its own "rightly guided" (i.e., faithful to God's word) communities working, studying, and praying together. Its goal was the establishment of a separate community of true believers in Egypt. As in the Muslim Brotherhood and Pakistan's Jamaat-i-Islami, there were gradations of membership. Full members were expected to devote themselves totally to the work of the community, leaving their jobs, family, and former friends behind. The ideal was martyrdom: a willingness to give up even one's life in the struggle for Islam. The Amir or leader demanded unquestioning obedience. Errant members might be excommunicated and/or punished.

While members of Muhammad's Youth were equally militant, well organized, and disciplined, they were governed more democratically by an executive council which relied on consultation rather than one-man rule. Thus the council had decided to seize the Military Academy and assassinate Sadat despite their leader's personal belief that such action was premature.

But who are these militants? Contrary to many stereotypes and expectations, they have not been uneducated peasants ignorant of the modern world, rejecting modernization in order to bury themselves in the past. The leadership of Muhammad's Youth and Takfir wal-Hijra combined an early traditional religious upbringing with a modern education. Dr. Salih Siriya, founder of Muhammad's Youth, had earned a doctorate in science education. Born in Palestine, he had been a member of the Islamic Liberation Organization (ILO), founded in Jordan by a former follower of Hassan al-Banna. The ILO is an international Islamic organization with a network of clandestine cells in many Muslim countries. It is very secretive, strongly anti-Western, particularly interested in recruiting members from the military, and dedicated to the restoration of the caliphate (a universal Islamic state) and the creation of a unified worldwide Islamic movement. Shukri Mustafa, founder of Takfir wal-Hijra, held a bachelor's degree in agricultural science. Both Mustafa and Siriya had been members of the Muslim Brotherhood and as a result had been imprisoned. Mustafa had in fact been imprisoned in Nasser's suppression of the Brotherhood in 1965 and was among those released in 1971 by Sadat. Though Siriya and Mustafa honored the memory of Hassan al-Banna and Sayyid Qutb, they believed that the Brotherhood had drifted from its early commitment, that its members were broken, mellowed, or burned out. Therefore each had begun to develop his own organization during the late sixties and early seventies.

The key leaders of Jamaat al-Jihad have been Muhammad al-Farag, Col. Abbud al-Zumur, Lt. Khalid Islambuli, and Shaykh Umar Abd al-Rahman, reflecting a civilian, military, and religious alliance. Farag, an electrician, was the chief ideologist of al-Jihad. Like the founders of other militant organizations, he was a former Muslim Brother who had become disaffected by its moderate posture. Col. al-Zumur and Lt. Islambuli were primarily responsible for Sadat's assassination.[23] Shaykh Abd al-Rahman, the religious adviser for al-Jihad, issued decrees which religiously justified al-Jihad's actions.

Profile of Membership

Activist organizations recruited members from local mosques, schools, and universities and were organized into secret cells. Many members were modern-educated, highly motivated individuals from the lower middle and middle class. The majority had university degrees in modern scientific and technical professions like engineering, medicine, science, and law rather than in religion or the humanities. Most had migrated from villages and towns to Cairo, Alexandria, and Asyut. One Egyptian expert concluded:

> The typical social profile of members of militant Islamic groups could be summarized as being young (early twenties), of rural or small-town background, from middle and lower middle class, with high achievement motivation, upwardly mobile, with science or engineering education, and from a normally cohesive family. It is sometimes assumed in social science that recruits of "radical movements" must be somehow alienated, marginal, anomic, or otherwise abnormal. Most of those we investigated would be considered model young Egyptians.[24]

However, these young Egyptians encountered vividly the "new ways" of city life. The wealth and ostentatious life-styles of the rich contrasted starkly with the poverty and massive unemployment of overcrowded ghettos, and the clash between Western and traditional Islamic values in dress and social (especially sexual) mores on the streets and in the media was a further source of scandal and outrage. While the freedom and delights of city life proved seductive for some, for many Islamically minded young Egyptians modern Egyptian life produced a sense of isolation and alienation.

Independent private mosques, freed from government control, have proven especially important in the development of Islamic movements, providing a center for organization and recruitment. They provide natural meeting places and organizational centers for Islamic militants. During the seventies the number of private mosques doubled from approximately twenty to forty thousand. Out of forty-six thousand mosques in Egypt, only six thousand were controlled by the Ministry of Religious Endowments. Private mosques and their preachers, in contrast to state-supported mosques, were financially and politically independent. The religious establishment and the government-controlled media were often seriously undermined by fiery sermons in private mosques delivered by charismatic preachers.

Groups like al-Jihad extended their influence through a network of educational and social welfare societies, including Quran study groups for the faithful and social centers which offered food, clothing, and assistance in obtaining housing to the poor in crowded urban areas. Student organizations at universities assisted with free books, clothes (including "Islamic dress" for women), tutoring, and housing. In the membership, Egypt's younger generation was heavily represented. Islamic organizations offered a new sense of community, a society organized into cells called families (*usrah*). More important, their community was based upon an Islamic ideology which provided a sense of identity and continuity as well as a critique of society and an agenda for radical change.

The educational and religious backgrounds of many militants remind us that, however repugnant the acts of extremists might be, they were not mindless or irrational. Specific causes motivated their actions: food riots in January 1977 in Cairo, unemployment, acute housing shortages, poverty, Sadat's open door (*al-infitah*) economic program, the Camp David Accords with Israel. Sadat's religious and secular critics charged that the open-door policy simply meant greater Western (especially American) economic involvement and dependence which lined the pockets of multinational companies and Egyptian elites but failed to address the basic economic and social problems of the majority of Egyptians. Camp David was viewed by many Egyptians and by most Arab and Muslim governments, which broke diplomatic ties with Egypt, as abandonment of the Palestinians displaced or living under Israeli occupation, and a unilateral capitulation to Israel and, by extension, to its American patron.

Egyptian Islamic Movements in the Nineties

Despite sporadic eruptions of violence and continued confrontation between the government and Islamic militants, in general a quiet rather than a violent revolution has occurred in Egypt. The impact of the Islamic resurgence may be witnessed across the religious, political, and socioeconomic spectrum of society—in the continued strength of conservative establishment Islam, the new vitality of Sufi mysticism, and the proliferation of moderate Islamically oriented organizations and voluntary social welfare associations.[25] The desire to lead a more Islamically informed way of life can be found among the middle and upper class, uneducated and educated, peasants and professionals, young and old, women and men. Mosques abound, built both by the government and by private individuals. They include not only formal buildings but thousands of structures and rooms added onto hotels, hospitals, and private dwellings. Religious programming and literature is not only more evident in the government-controlled media and in newspapers, but also in bookshops and scattered among the popular secular magazines and books of street vendors. As previously noted, preachers such as Shaykh Muhammad Mitwali al-Shaarawi and Abd al-Hamid Kishk, an outspoken government critic often regarded as an extremist and imprisoned by both Nasser and Sadat, have become religious media stars in Egypt and the Arab world. Their voices are heard not only in mosques and religious gatherings but also on cassettes played in taxicabs, in shops, on the streets, and in the homes of the poor and the middle class alike.

Egypt's militant groups remain fragmented and small. They do not enjoy broad popular support, though at times their critique of society has won admiration or support in a society often on the brink of economic disaster. However, social discontent has not always translated into membership. Thus they remain a distinct minority voice in society. The Muslim Brotherhood and a host of smaller Islamic societies and organizations enjoy more support and function as effective voices for social and political change. Some are political, while others remain apolitical religious and social activists. While the Muslim Brotherhood has grown and become a major voice in Egyptian life, again one must remember that its message is not the dominant one but that of a significant, well-organized, vocal minority.

Lebanon's AMAL and Hizbullah

Just as Iran and the Ayatollah Khomeini became household names during the eighties, so too Lebanon's AMAL and Hizbullah became part and parcel of the evening news headlines. As in Iran, Shii Islam in Lebanon became the basis for the political mobilization of Shii Muslims into protest and revolutionary movements. Lebanon often seemed another example of radical fundamentalism and of "Americans held hostage." As many in the West were stunned by the overthrow of the Shah of Iran and the rampant anti-Americanism in what had long been regarded as a strong Western ally, so too the disintegration of Lebanon— the "Switzerland of the Middle East," where Europeans and Gulf shayks invested their money and built summer palaces—was impossible to fathom. That both Western-oriented Iran and Lebanon would become hotbeds of Shii radicalism seemed irrational. Both Iranian and Lebanese leaders and those of the Western world appeared to have been caught completely off guard. As with other Islamic movements, however, the rise of Shii political activism in Lebanon must be seen within the context of Lebanese politics and society.

For much of Lebanon's modern history, Christians were the dominant political and economic power.[26] Muslims regarded themselves as second-class citizens, and Shii Muslims in particular were at the bottom of the ladder of privilege. But in the seventies and eighties the status quo was challenged as Shii political organizations and militias sprang up to vie with Christian, Israeli, and Palestinian military forces. As in Iran, the emergence of Islam in Lebanese politics was the product of sociopolitical factors combined with a charismatic leadership that skillfully used religious symbols to organize popular support. However, in contrast to Iran, Shii Islamic movements grew in a soil in which the Shii community was a distinct religious minority, predominantly rural, poor, disorganized, and lacking an effective clerical leadership.

Lebanon has often been viewed as another modern, Western-oriented casualty of militant Islam. Lebanon's modern, Westernized capital of Beirut had been a banking, financial, educational, and cultural center of the Middle East. Its beaches and mountain resorts attracted great waves of tourists from Europe and the Middle East. Foreign presence and influence could be found everywhere in its restaurants, stores, boutiques, nightclubs, schools and universities, fashions, cinema, books, and high-tech products from the United States, Europe and Asia. However, this prosperous country, an ally of the West, would be torn apart by a fifteen-year civil war. Lebanon became a land of death and unimaginable destruction, a country without a representative government, a fertile ground for the development of militant Shii politics and Iranian influence, a battleground of Muslim and Christian militias, Israeli and Palestinian forces, a land of hijackings, kidnappings, and attacks on Western embassies and personnel.

Lebanon is an example of colonialism's mixed legacy. U.S. and European influence could be seen in its history. Once under Ottoman rule, Lebanon had become a French protectorate during the mandate period. The modern state of Lebanon, which includes what was formerly Greater Syria, is a nation whose geographic boundaries were artificially determined by the French. Long regarded as patrons and protectors of the Christian (especially Maronite) communities, the French assured a Christian-dominated state. Beirut became a center of American and European influence, symbolized by large foreign diplomatic and business communities, American-, French-, and British-run schools and banks, and a Western-educated, trilingual (Arabic, French, and English) elite class. Indeed, to have obtained the best education meant that, whether Christian or Muslim, a Lebanese had attended American or European primary and secondary schools often run by Christian missionaries and then studied at the American University of Beirut (formerly the Syrian Protestant College) or the French Jesuit University of Saint-Joseph.

Lebanon's multiconfessional government, like the geographic boundaries of the state, reflected the politics of French colonialism and the realities of religious communalism. The state had been created in such a way as to assure Christian dominance under French protection. Religion has played a real though at times amorphous role in Lebanese society.

Augustus Richard Norton has observed: "Religion is an ascriptive fact in Lebanon. The Lebanese is born into a religion, which conditions his cultural realm and that decisively defines his social and political identity. Thus, religion as a label is an inescapable fact, both because of Lebanon's natural history and her political system that only takes account of citizens as members of one religion or another."[27]

Based upon a French-conducted census in 1932, a National Pact or Covenant was drawn up in 1943, assigning political power and representation on a proportional basis to the major religious/confessional communities: Maronite Christian, Sunni Muslim, Shii Muslim, and Druze.[28] Thus the president was a Christian, the prime minister a Sunni Muslim, and the speaker of the Chamber of Deputies a Shii Muslim. Key positions in the cabinet, parliament, bureaucracy, and military were distributed proportionally among the confessional groups.

The Shii (Shia) of Lebanon were politically disorganized and the most economically disadvantaged of the communities. Two imams (religious leaders) from Iran—Imam Musa Sadr and Imam Khomeini—and the establishment of the two organizations that they inspired, AMAL and the Hizbullah, reversed that situation.

The Revolt of the Dispossessed: Musa Sadr and AMAL

If the Shii of Lebanon had been leaderless and disorganized, by the eighties they possessed a broadly based political organization and militia. Much of the resurrection of the Shii community is due to the charismatic leadership of Musa Sadr, the founder of AMAL. Musa Sadr came from a prominent clerical family. Of Lebanese ancestry, he was born in Iran and educated in Najaf, Iraq, a major center of Islamic learning. Many of Shii Islam's leading ayatollahs and scholars were based in Najaf. While most were apolitical, Najaf would also come to be remembered for two militant voices, Muhammad Baqir al-Sadr, Musa Sadr's cousin, who was executed by Saddam Hussein, and the Ayatollah Khomeini, who taught there during part of his thirteen-year (1965–78) exile.

There had been centuries of close religious and cultural ties between the Shii of Lebanon and those of Iran and Iraq. The latter countries were centers of clerical training and pilgrimage, containing the shrines and tombs of the family of Ali and many great Shii leaders. Musa Sadr's family reflected this connection. It had originally come from Lebanon, so he accepted an invitation to "return" to lead the Shii community in Tyre in southern Lebanon. With the exception of a few wealthy and powerful Shii feudal families, the Shii of Lebanon, who were concentrated in the south, Beirut, and the Bika valley, trailed the other communities politically and socioeconomically. But within little more than a decade, Musa Sadr asserted his role as a major Shii leader and brought organization and a sense of communal identity and purpose to the Shii of Lebanon. He became leader of the Supreme Islamic Shii Council, a group of prominent Shii religious leaders, and founder of the Movement for the Dispossessed, a religiosocial movement whose goal was parity and social justice.

Although Lebanon's demographics had changed and Muslims (Sunni and Shii) were now a majority, Maronites continued to dominate the government. The Shii, who had grown from 18 percent of the total population in 1932 to 30 percent in 1968, by the nineties constituted more than one third of Lebanon's population (one million out of a total population of two and one half to three million). Despite changes in demography which made them Lebanon's largest community, however, the Shii remained a distant third in political and economic power. In the midst of the wealth and conspicuous consumption of a prosperous Beirut, sprawling Shii slums expanded. Moreover, "the Shiis constituted the highest percentage (twenty-two percent) of families earning less than 1,500 Lebanese pounds. According to every indicator, the Shiis were at the very bottom of the socio-economic ladder."[29]

The Movement for the Dispossessed provided Musa Sadr with a platform and vehicle to mobilize Lebanese Shii and to pressure the government for socioeconomic reforms. As in Iran, Shii history and belief were interpreted to provide an ideology of protest against social injustice and to champion the rights of the disinherited and oppressed. Early Shii suffering at the hands of Sunni rulers, in particular the martyrdom of the revered Shii Imam Husain by the army of the caliph Yazid at the battle of Karbala in 680, were equated with the discrimination and exploitation suffered by Shii under the Christian-dominated confessional system. At times Sadr's words and images could belie his chosen role as a reformer rather than a violent revolutionary:

> This revolution did not die in the sands of Karbala, it flowed into the lifestream of the Islamic world and passed from generation to generation, even to our day. . . . It is a deposit placed in our hand that we may profit from it as from a source of new reform, a new position, a new movement, a new revolution to repel the darkness, to stop tyranny, and to pulverize evil. Brothers, line up in the row of your choice: that of tyranny or that of Hussein [Husain]. I am certain that you will not choose anything but the row of revolution and martyrdom for the realization of justice and the destruction of tyranny.[30]

Mass demonstrations, strikes, and marches were organized to press Shii demands for a more equitable reapportionment of political representation and economic power, as well as improved housing, hospitals, roads, irrigation projects, and educational opportunities. Sadr found ready recruits. The Shii, unlike the Christians, Sunni, and Druze, had no political parties. An emerging new Shii middle class of merchants and young professionals who had taken advantage of rapid modernization, along with university students, had become politicized and involved in multiconfessional leftist and communist groups. They now found a Shii alternative for social protest under a dynamic and charismatic leader.

In 1975, with civil war impending, Musa Sadr created the militia AMAL to protect Shii rights and interests. AMAL is an acronym for the Lebanese Resistance Battalions (Afwaj al-Muqawimah al-Lubnaniyah) which also means "hope." In a land of militias, the Shii had been the only major community without one. Although initially AMAL was a militia or military wing of the Movement for the Dispossessed, it absorbed the sociopolitical movement within it. Socioeconomic disparities along with several major events contributed to the growth of AMAL and the radicalization of Shii in Lebanon: the Lebanese civil war of 1975–76, the disappearance of Musa Sadr in 1978, the Iranian Revolution of 1979, and Israeli invasions and occupation of Lebanon in 1978 and 1982.

The outbreak of Lebanese civil war in 1975 exacerbated communal differences, divided communities, undermined interconfessional alliances, and drove most Lebanese to fall back upon their sectarian/confessional community for protection. Despite the war, the Maronite-dominated government refused to address the new demographic realities. Shii grievances and demands for a redistribution of power went unheeded, contributing to further radicalization and the belief that the Lebanese system was no longer a representative government but simply Christian minority rule.

Next, Shii in southern Lebanon were caught in the crossfire as their lands became a battleground between entrenched PLO forces and Israeli commandos. The Israeli invasion of southern Lebanon in 1978 realized the worst Shii fears with "an active campaign of air attacks, raids, kidnappings and house bombings."[31]

The mysterious disappearance of Musa Sadr in 1978, while on a visit to Libya to meet with Muammar Qaddafi, marked a turning point in the history of AMAL, profoundly affecting its growth and leadership. AMAL now had its martyr and Imam. However, his influence encompassed all the disparate elements within the Shii community of Lebanon.

In Shia Lebanon, in the slums of Greater Beirut, in the villages of the Bekaa Valley, and in the south, posters bearing Imam al Sadr's picture were everywhere. He was claimed by sly politicians, by starry-eyed young boys, pushed out of the villages of the south into the anarchy and confusion of Beirut. . . . The Shia were to swing all the way from quietism to martyrology. Men needed "saints," and in Sayyid Musa they found the elements out of which militant sainthood could be constructed. His aura hovered over the ruined world of the Shia in Lebanon, and its politics became, in many ways, a fight over the realm of a vanished Imam.[32]

Eclipsed by the civil war, AMAL now took on new life. Although many believed that Musa Sadr died in Libya, his disappearance was easily associated with Shii doctrines of martyrdom and of the occultation or seclusion of the "Hidden Imam" (i.e., the last Shii Imam who disappeared in the ninth century and whose return is awaited).[33] Like the Ayatollah Khomeini, Musa Sadr had cultivated a religious persona, often identifying his role with the spiritual leadership of the great early Imams, Ali and Husain. Nor did he discourage those who called him "Imam," a title which, strictly speaking, only belongs to the successor of Muhammad, through Ali, as religiopolitical leader of Islam. Like Khomeini, he became a symbol of Shii political activism—in death, even more of a cult hero, the symbol of martyrdom and messianic hope.

The Iranian Revolution, in Lebanon as in much of the Muslim world, was a source of inspiration and awe. The discrediting of Qaddafi caused by the disappearance of Musa Sadr, coupled with Iran's "Islamic" revolution, attracted many Shii businessmen and professionals away from Arab nationalism and the Left and toward AMAL. Ironically, it also led to the transformation of its leadership from clerical to a more secular orientation as Nabih Berri, a lawyer and representative of the new Shii professional class, gained control of AMAL. By 1982 Berri was at the head of an AMAL that was clearly the largest Shii political organization or militia and a formidable political force in Lebanese politics. However, despite ties with Iran, Berri steered AMAL on a nationalist path. AMAL's primary identity was as a Shii nationalist organization. It sought political and economic parity for the Shii community, working within the multiconfessional political system; thus its goal was reform, not revolution. In contrast to Iran and to Hizbullah, it did not seek to establish an Islamic state.

The Rise of Hizbullah

The impact of the Iranian revolution, the Israeli invasion of Lebanon in 1982, Israel's subsequent occupation of the south (1982–85), and the massacre of Palestinians and Lebanese in the Shatila and Sabra refugee camps by Christian Phalangists with Israeli complicity, contributed to the formation of extremist revolutionary Islamic organizations, in particular Hizbullah (Party of God) and al-Jihad, whose goals were to emulate Iran. Following Iran's example, they were against the United States and Israel and sought to violently overthrow the Lebanese government and impose an Islamic state. Hizbullah's pro-Iranian orientation was clearly set forth in its manifesto:

> We, the sons of Hizbullah's nation, whose vanguard God has given victory in Iran and which has established the nucleus of the world's central Islamic state, abide by the orders of a single wise and just command currently embodied in the supreme Ayatollah Khumaini. . . . We have opted for religion, freedom and dignity over humiliation and constant submission to America and its allies and Zionism and their [Christian] Phalangist allies. We have risen to liberate our country, to drive the imperialists and the invaders out of it and to take our fate in our hands.[34]

The U.S. relationship with Lebanon had spanned decades, involving an educational, missionary, commercial, and diplomatic presence; in 1958 President Eisenhower had sent Marines to protect Lebanon's president. Yet the United States had supplied Israel with aircraft and arms and failed to condemn Israel's 1982 invasion and occupation of Lebanon in the United Nations. Many Lebanese regarded the United States as a tacit partner in Israel's sweep through Lebanon, and it had continued to support Lebanon's (Maronite) Pres. Amin Gemayel, whose regime radical Shii regarded as unrepresentative and illegitimate. Thus for many in 1958 the multinational force was seen as part of the problem and not the solution—as protecting the Gemayel government and Israeli interests.

The year 1982 proved to be a turning point in Shii politics. In addition to the Israeli invasion, it witnessed a schism in AMAL, the radicalization of Shii populist movements and politics, and with it a major challenge to AMAL's dominance of Shii political activism. When Nabih Berri agreed to participate in the National Salvation Committee, an alliance with other confessional groups, shortly after the Israeli invasion of 1982, Hussein Musawi, a former teacher and member of AMAL's Command Council, challenged his leadership. Musawi accused Berri of collaboration with Israel and rejected his secular nationalist goal of a multiconfessional state. He broke away and withdrew to Baalbek in the Bika Valley where he formed Islamic AMAL, a militant organization strongly influenced by Khomeini's Iran and committed to the creation of an Islamic republic in Lebanon. Baalbek was already a center of Iranian and militant Shii activity. Iran had sent fifteen hundred Revolutionary Guards (Pasdaran) to Lebanon in 1982 shortly after the Israeli invasion, to assist in the resistance. There they provided training and support for Hizbullah, an Iranian-inspired radical Shii organization which had emerged in the wake of the Iranian revolution, but which came to prominence only as a response to the 1982 Israeli invasion.[35]

In contrast to AMAL and Islamic AMAL, Hizbullah's leadership was clerical and mosque-based. Young pro-Iranian Lebanese Shii clerics led groups of young people and professionals in a network of loosely organized militias. Many of these clerics had been educated in Najaf, Iraq, during the sixties and seventies, when it was a center not only of Islamic learning but also of militant Islamic thought. Among its leading Iraqi ayatollahs were Muhsin al-Hakim and Baqir al-Sadr, who were joined in 1965 by the Ayatollah Khomeini. It was here that Lebanese students, sitting beside their colleagues who had come from all over the Shii Muslim world to study at Najaf's seminary, imbibed an ideological worldview that countered the inroads of the West and posited Islam as an alternative to nationalism and communism. Senior clerics lectured and wrote on such topics as Islamic government, law, economics, and revolution. It was in Najaf in 1970 that the Ayatollah Khomeini delivered his lectures on the nature of Islamic government and clerical rule. Though they received little recognition at the time, by 1979 an idea had been transformed into a reality in postrevolutionary Iran and become a source of inspiration and guidance for many Shii in Lebanon.

For the young Lebanese Shii clerics and their followers, Khomeini and Iran became spiritual and financial godfathers. Posters and pictures of the Ayatollah Khomeini were to be found in homes and mosques, on public walls, and at marches and rallies. His speeches and writings were the source of sermons and political speeches. As Musa Sadr's picture came to dominate the AMAL newspaper's masthead, so the visage of Khomeini graced the masthead of Hizbullah's *Covenant*. Much as in Iran, women wore the *hijab* (head scarf) and men sprouted beards, considered a sign of Islamic identity, in emulation of the Prophet. Hizbullah leaders traveled regularly to Iran, as did Iranian officials to Lebanon. The Iranian embassy in Damascus (Syria was one of the few Arab countries that was an ally of Iran) became a center for organizing and coordinating Iranian support (training, money, and weapons) for political violence. Iran provided training and financial support for Hizbullah's military operations and social services. By 1987, estimates of Iranian aid were as high as ten million dollars per month. Iranian materials were used to train and indoctrinate a new generation of holy warriors. At the same time Hizbullah won the hearts and minds of new supporters and retained the loyalty of others through an impressive program of social assistance which included housing, schools, hospitals, clinics, cooperative supermarkets, and financial assistance and educational scholarships for the victims of war. Some reports estimated that Hizbullah provided scholarships for some forty thousand students.[36]

Organization and Leadership of Hizbullah

Organizationally, Hizbullah was more of a loosely organized ideological movement than a political party. In contrast to AMAL's political organization, identification with a single leader, and fixed membership, Hizbullah was a confederation of like-minded groups and militias spread out in the neighborhoods, villages, and towns of Beirut, the Bika Valley, and southern Lebanon. By the mideighties Hizbullah encompassed a number of revolutionary groups: Islamic AMAL, Jund Allah (Army of God), the Hussein Death Squad, the Revolutionary Justice Organization, and al-Jihad. All shared a common outlook and agenda: the dismantling of the Lebanese state and the creation in its place of an Islamic state; the acceptance of Khomeini and Iran as the model to be emulated; a consensus that its enemies were the Lebanese government, other confessional groups, the United States, and France, as well as pro-Western Arab governments such as Saudi Arabia and Kuwait; and belief that it was a religious duty to destroy the "enemies of God" through jihad, martyrdom, and self-sacrifice. There were many leaders, rather than one. While they cooperated in joint ventures, they also felt free to act on their own. Hizbullah became more structured, with a Supreme Consultative Council of clergy and militia leaders as well as regional councils and committees. However, it remained more of an umbrella group than a tightly knit movement.

> Much ink has been spilt on the issue of who leads Hizbullah, but the fact is that—at best—we are dealing with a collegial leadership which subsumes many factions and cliques. Even Iran must cajole rather than direct and order in such an environment. In other words, despite their links to Iran, it would be erroneous to assert that the cliques within Hizbullah lack a considerable freedom of action.[37]

Over the years a number of senior leaders have emerged, in particular shaykhs Abbas al-Musawi (killed in 1992), Subhi al-Tufayli, Raghib Harb, and especially Muhammad Husayn Fadlallah. Shaykh Fadlallah (b. 1935) is perhaps the most prominent Shii cleric in Lebanon, a religious scholar whose reputation and following extend beyond Lebanon. Although he has resolutely maintained that he is not the leader or even a member of Hizbullah, he is widely regarded as its spiritual guide.[38] Born and educated in Najaf, in 1966 he returned to Lebanon, his parents' homeland, where he quickly established his reputation as a leading religious authority. A gifted preacher, he draws large crowds to his mosque in Beirut. Although he is primarily a religious scholar, the experience of the Lebanese civil war and the Iranian revolution influenced his espousal of political activism.[39] Like the Ayatollah Khomeini, Musa Sadr, and other Shii activist leaders, Fadlallah combines traditional religious scholarship with a powerful reinterpretation of Islamic history and belief that emphasizes political activism and social reform: "As Moslems we consider politics to be part of our whole life, because the Koran emphasizes the establishment of justice as a divine mission. . . . In this sense, the politics of the faithful is a kind of prayer."[40] Thus while he has written a multivolume commentary on the Quran, he has also authored *Islam and the Logic of Force* and *The Islamic Resistance*. His message rejects quietism in the face of the oppressed status of Shii life and advocates struggle against social injustice: "Muslims are to embark on an 'Islamic revolution' . . . under the guidance of religious officials, the 'ulama,' whose knowledge and integrity guarantee the ultimate triumph of Islam over the Satanic forces of disbelief."[41]

Fadlallah is an activist but denies any formal political role: "I am a servant of the community, working for Moslems and non-Moslems. I even have supporters among the Sunnis because I speak about Islam as a movement to establish freedom and justice."[42] His counsel is sought, although not always followed, and he often speaks out in defense of Hizbullah and related groups. Where contradictions or discrepancies exist between what Fadlallah says and what Hizbullah does, it is difficult to know how much this is due to genuine differences as opposed to Fadlallah's disingenuousness.

Ideologically, Hizbullah's Islamic revolutionary orientation differs from the more reformist, nationalist stance of AMAL. While Hizbullah demands an Islamic state, AMAL has continued to respect the integrity of the Lebanese state and worked toward a multiconfessional state with a more equitable distribution of power. Hizbullah's clerical leadership, high Islamic profile, close Iranian ties, and near cult of the "path of the Imam Khomeini" contrasts with AMAL's nonclerical, more Westernized and moderate image. While Hizbullah assumes a more universal, international, and self-consciously Islamic orientation, AMAL is content to focus almost solely on national interests. Hizbullah's name, the Party of God, indicates its transnational identity as a movement of all Muslims that extends beyond Lebanon. Like Khomeini's movement in

Iran, it sees itself as part of a worldwide Islamic revolution and champions Pan-Islamic causes such as the liberation of Palestine.[43] In contrasting Hizbullah's and AMAL's relations with the Palestinians, Shaykh Fadlallah's comments reveal the difference between Hizbullah's more worldwide or Pan-Islamic concerns and AMAL's narrower nationalist commitment: "Hizbullah's strategy is a strategy of jihad that insists the presence of Israel in Palestine is illegal . . . as for AMAL, it may consider the liberation of the south [of Lebanon] as its sole task."[44]

Conflict Between Hizbullah and AMAL

Like activists in Iran, Hizbullah also employed Shii doctrines of self-sacrifice and martyrdom in the service of a jihad against oppression and injustice. This at times took the startling form of suicide and car bomb attacks against Western and Israeli targets. A series of suicide bombings were aimed at the American embassy, French and American military compounds, and businesses during 1983–84. More than 250 U.S. Marines were killed in the October 1983 bombing of their barracks on the outskirts of Beirut. Israeli forces were similarly targeted from 1983 to 1985. The withdrawal of the Multinational Force from Beirut and the retreat of Israeli forces to their security zone in southern Lebanon enhanced Hizbullah's image as the most effective defender of Shii rights and interests.

Hizbullah's use of force and its successes constantly put pressure on AMAL, with whom it increasingly competed for the minds and hearts of Shii, to forsake moderation and prove its militancy and effectiveness. The split in AMAL in 1982 and the growing prominence of radical organizations like Hizbullah eroded AMAL's primacy as defender of Shii interests. The two groups' ideological and political differences led to confrontations and armed conflict as they often fought each other rather than a common enemy.

AMAL differed with Hizbullah and related groups in its attitude toward the West and Israel. Berri traveled regularly to the United States, where he held a green card as a resident alien and his ex-wife and children resided. AMAL's policy was more focused in its concerns about U.S. policy and Israeli intervention, which contrasted sharply with Hizbullah's unrestrained anti-Americanism and militant cries of "Death to America." AMAL clashed with Islamic radicals over the TWA 847 hijacking in 1985, kidnapping, and hostage taking, and especially the kidnapping and execution of Col. William R. Higgins of the U.N. peacekeeping force (UNIFIL) in Lebanon in 1989. Berri had worked for the release of the TWA passengers. In contrast to Hizbullah, which viewed the U.N. presence as another form of foreign (American and Israeli) intervention, AMAL saw it as necessary to restore stability in southern Lebanon and sought to guarantee its security. Higgins's abduction triggered a fierce battle between AMAL and Hizbullah militias for control of West Beirut and southern Lebanon. Though Syria was an ally of Iran, it increasingly backed AMAL because Hizbullah's call for an Islamic state threatened Syrian preeminence in Lebanon.

The differences between AMAL and Hizbullah continued right up to the very end of the Lebanese civil war. Whereas AMAL joined with other confessional groups in accepting the Saudi-brokered Taif Accords, which provided a more equitable distribution of power (while retaining a Maronite president), Hizbullah remained adamant in its rejection of the accords. Hizbullah's strength was dramatically affected by the restoration of peace in Lebanon and a change in its relationship with Iran. With the Ayatollah Khomeini's death and the end of the Iran–Iraq war, Iran's government altered its priorities, emphasizing domestic development and international trade. Iran's new president, Hashemi Rafsanjani, pursued a more pragmatic and less militant foreign policy, halting support for the violent export of its revolution, and denounced the taking of hostages. Hizbullah's aid from Iran was cut dramatically and its isolation increased. At the same time Syria pressured Hizbullah to release its hostages. With the end of the Cold War and the loss of Soviet patronage, Syria sought to shed its terrorist image and improve its relations with the West.

As in Egypt, sociopolitical realities had spawned a number of Islamic movements in Lebanon. Among the Shii the disparities between rich and poor, and the marginalization of the Shii masses, proved fertile ground for both reformist and revolutionary movements. Ironically, Iranians influenced the formation of both AMAL and Hizbullah. The former owed its origins and inspiration to Musa Sadr and the latter was ideologically and financially indebted to the Ayatollah Khomeini. Yet despite their common soil, AMAL pursued a path of reform within the established state while Hizbullah rejected the past and sought to impose its vision of an Islamic order. While AMAL, though not adverse to armed conflict, was careful to foster relations with the West, Hizbullah rejected the "Great Satan" and boldly pressed its claims through violence and terrorism.

Islamic Politics in the Maghreb

For much of the eighties, while world attention was riveted on Iran and Libya, Khomeini and Qaddafi, few studies of Islamic revivalism took

North Africa seriously. Indeed, many would have observed, "It may have occurred in the Middle East but never in francophone, if not Francophile, North Africa: Algeria, Tunisia, and Morocco." Their authoritarian regimes banned Islamic parties and political organizations. While it acknowledged Islam in its constitution, Algeria was basically regarded as a hard-line socialist state, firmly controlled by a secular-oriented leadership. Tunisia under Habib Bourguiba was modern and Western-oriented. In his pursuit of modernization, Bourguiba had publicly broken the fast of Ramadan and outlawed (rather than simply restricted) polygamy. Morocco's leader, King Hassan, appeared to be successful in combining a religious dimension to his political legitimacy with a strong Western orientation and tight control over any and all opposition.

By the end of the eighties, however, electoral politics indicated that the leading political opposition in Tunisia and Algeria was Islamic: the Islamic Tendency Movement (MTI) or Renaissance Party (Ennahda) in Tunisia and the Islamic Salvation Front in Algeria. Though prevented from running as a political party, members of MTI won 17 percent, and in some urban areas as much as 40 percent, of the vote in government-controlled national elections in Tunisia in April 1989. Algeria and, indeed, much of the Muslim world seemed traumatized as the Islamic Salvation Front, within the space of only a year, went from being an outlawed party to scoring a stunning upset and sweeping the municipal elections in 1990 and then subsequently the parliamentary elections in 1991.

Islamic movements in the Maghreb, as in most of the Muslim world, have developed a new or modern Islamic form of organization. In contrast to traditional religious organizations, Sufi mystical brotherhoods (*tariqah*) and *ulama* associations active in the nationalist struggle, modern Islamic activist organizations have a lay rather than clerical leadership, are urban-based, and thrive primarily among students and educated professionals while also attracting members of traditional occupations (merchants, traders, craftsmen). In their formative stage, most of these organizations have been particularly influenced by the trailblazers of contemporary Islamic activism: Mawlana Mawdudi, Hassan al-Banna, and Sayyid Qutb. In addition they draw on early Islamic reformers like ibn-Hazm and ibn-Taymiyya as well as twentieth-century Muslim activists and nationalist leaders in the Maghreb such as Malek Bennabi and Ben Badis. Like their counterparts in other parts of the Muslim world, in recent years, where governments permit, Islamic activists have moved from the periphery to the center and now participate in electoral politics and mainstream society. Most advocate working within the system rather than using violence to come to power.

The Maghreb also offers an excellent example of the evolution of many

Islamic movements. From relatively small apolitical Quran or religio-cultural reform societies, Islamic movements there have come to function as modern social movements and participatory political parties or organizations that espouse populist causes like political liberalization, jobs, better services, integrity in government, human rights, more equitable distribution of wealth, and greater emphasis on Arab–Islamic cultural identity. They now include both a first and a second generation of members who are educated and middle-class, organizationally and intellectually sophisticated. At the same time both leadership and rank and file include a diversity of religious, political, and educational orientations—conservative and reformist, moderate and militant. The case of Tunisia's Islamic Tendency Movement shows the nature of Islamic movements in the Maghreb and the dangers of radicalization.

Tunisia and the Islamic Movement

For most of Tunisia's history from its independence in 1956 to 1987, it had but one ruler. Habib Bourguiba, the "Great Combatant" of Tunisia's nationalist struggle, ruled a unified, one-party political state for more than thirty years. More than any other Muslim ruler except Turkey's Ataturk, who established a totally secular state, Bourguiba set Tunisia upon a path of modernization which was decidedly pro-Western and secular, becoming a valued friend of both France and the United States. Tunisia's Arab–Islamic heritage was overshadowed by an official and elite Francophile culture. French rather than Arabic was the official language of government, higher education, and elite culture and society. Bourguiba carefully circumscribed the presence and influence of Islam. Shortly after independence, Tunisia passed a Personal Status Law (1957) which went further than any other Muslim country except secular Turkey in banning rather than simply restricting polygamy. Even more symbolic of Bourguiba's approach to religion and modernization and his wholehearted acceptance of Western values, was the abolition of Sharia courts, the ban on the wearing of the *hijab* (head scarf) by women, and his attempt to get workers to ignore the fast of Ramadan, during which he drank a glass of orange juice on national television. He decried the deleterious effects of fasting during daylight hours, claiming that it affected national productivity and economic development. The Zaytouna, a famed center of Islamic learning in North Africa and the Muslim world, was closed. The *ulama* were debilitated rather than, as in many Muslim countries, coopted by the government. For Bourguiba, Islam represented the past; the West was Tunisia's only hope for a modern future. No wonder, then, that experts concluded: "Historically the most open and Mediterranean of the Arab countries, Tunisia is an improbable site for an Islamist upsurge."[45]

The development of Islamic movements as a reaction to personal lives, social conditions, and government policies and actions can be seen quite clearly at every stage in the development (from its formation through its major transformations) of Tunisia's major Islamic organization, the Islamic Tendency Movement or, as it is known today, the Renaissance Party (Hizb al-Nahda or Ennahda).

The life of its paramount leader, Rashid Ghannoushi, follows the pattern of many activists of his generation:

1. Traditional and then modern education, but with a clear emphasis on Arabic rather than French.

2. Youthful enthusiasm and participation in the Arab nationalist movement of Gamal Abdel Nasser.

3. Subsequent disillusionment and an identity crisis when Arab nationalism was discredited by the bitter Arab defeat in the 1967 war.

4. A turn away from Arab nationalism and return to Islam, in particular an ideological revivalist worldview indebted to Hassan al-Banna, Sayyid Qutb, and Mawlana Mawdudi.

5. A strongly negative reaction to his direct encounter with Western culture and values, and a deeper reappropriation of his Arab–Islamic heritage after studying in France.

6. Assumption of the dual leadership role as teacher–activist, a path similar to that of many other activists like al-Banna and Qutb in Egypt, Mawdudi in Pakistan, and modern-day leaders such as Abdesallem Yassin in Morocco, Abbasi Madani in Algeria, and Hassan Turabi in the Sudan.

7. Subsequent realization of the limited relevance of outside activists, whose outlook and strategies were conditioned by their local experiences, and thus the need to contextualize the Islamic message and ideology.

8. A resulting transition from an Egyptian Muslim Brotherhood orientation to a more Tunisian-centered movement, and from a militant Qutb-inspired activism to more pragmatic, moderate, accommodationist activism.

Ghannoushi's development and that of the Islamic movement in Tunisia were shaped by events in the early seventies. The causes or influences were many: fallout from the defeat of 1967 and the discrediting of Arab nationalist ideas; high unemployment and food shortages resulting from the failure of the government's planned socialist economic development program (in particular the collectivization of agricultural lands from 1962 to 1970); the national strikes and food riots of 1978; the reemergence of Islam and of Islamic organizations such as the Egyptian Muslim Brotherhood and the Jamaat-i-Islami of Pakistan in Muslim politics; the use of oil revenues by Saudi Arabia and Libya to promote their influence in the Muslim world; and the Iranian Revolution.

The Tunisian government, anxious to distance itself from the failures of its socialist policies of the sixties, embarked upon a policy of political and economic liberalization in the seventies and also showed greater interest in the revivification of Tunisia's Arab–Islamic identity and heritage. The reassertion of this identity was reflected in the establishment of cultural societies, in particular the Quran Preservation Society at Tunisia's venerable Zaytouna mosque, and in the growth of Islamically oriented student groups at the universities and secondary schools, which contributed to the growth of Islamic revivalism. The reaffirmation of Islamic identity was also strongly influenced ideologically by Egypt's Muslim Brotherhood. The writings of Muslim Brothers such as Hassan al-Banna, Sayyid Qutb, Muhammad Qutb, and Muhammad al-Ghazzali were readily available and had a powerful influence on the nascent Islamic movement in Tunisia.

Rashid Ghannoushi proved a popular preacher, teacher, and youth leader. He drew large crowds from among the poor working class to his sermons as he moved from one mosque to another and attracted students from his school to the weekly discussions conducted at the Quran Preservation Society. The Society was dedicated to the promotion of Tunisia's Arab–Islamic heritage and enjoyed government sponsorship. To the extent that it remained an apolitical cultural society, Bourguiba found it useful as a foil against his leftist critics.

For Ghannoushi, teaching, preaching, and participation in the Society were interrelated aspects of his mission to restore Tunisia's Arab–Islamic roots and civilization. His experience and that of many others not from the elite classes reflected Tunisia's deep-seated identity crisis and the source of its weakness: "I remember we used to feel like strangers in our own country. We had been educated as Muslims and Arabs, while we could see that the country had been totally molded in the French cultural identity. For us the doors to any further education were closed since the university was completely Westernized. At that time, those wanting to continue their studies in Arabic had to go to the Middle East."[46]

Ghannoushi's preaching and writing drew heavily upon the interpretations and ideological worldview of Muslim scholar-activists such as Mawlana Mawdudi, Hassan al-Banna, Sayyid Qutb and his brother Muhammad Qutb, Malek Bennabi, and Muhammad Iqbal. The message was one condemning the political, social, economic, and cultural backwardness of Tunisian society, its loss of identity and morals owing to dependence upon a morally bankrupt and crisis-ridden Western society, and the need for a return to Islam. The culprits were the West, the Bourguiba government, and Tunisia's Westernized elites as well as the traditional religious establishment, which, coopted by the government, preached a stagnant rather than a dynamic Islam. Ghannoushi believed that the only hope for Tunisia, the Muslim world, and indeed the Third World was Islam. However, during the period 1970–78 the Islamic movement and its message remained primarily concerned with religiocultural change in society. Ghannoushi and a small group of colleagues, among them Abdelfattah Mourou, a lawyer and activist, pursued their goals within the context of the Quran Preservation Society. As an officially apolitical religiocultural movement, it posed little direct threat to the government. The official discussions and activities of the Society obscured the political consciousness and intellectual politicization that were also taking place among Ghannoushi and his colleagues.

The late seventies saw a progressive politicization of the Islamic movement in response to the internal situation in Tunisia and events in Iran. As in much of the Muslim world, the period 1978–79 proved to be a political turning point in the history of Tunisia's Islamic movement. Bourguiba's use of the military to brutally crush protest demonstrations in January 1978 and the subsequent victory of Iran's Islamic Revolution in 1979 underscored the failures of Tunisian Westernized secular society and fired the enthusiasm of many for a return to Islam. The confrontation between the government and the General Union of Tunisian Workers (UGTT) culminated in a general strike on January 26, 1978, in which many were killed or injured by government forces. Tunisia's trade union movement was considered by some to be the strongest in Africa and the Arab world, but Ghannoushi and the Islamic movement had avoided any involvement in it because they were leery of unionism, and because the UGTT was controlled first by the government's Destour Party and then by Marxists: "The social confrontation between rich and poor is a Marxist formula that did not correspond to our understanding of life. Later on, we realized that Islam also has a say in the confrontation, and that, as Muslims, we could not stay indifferent to it. Islam gives support to the oppressed."[47] The Islamists found themselves forced to rethink their ideas and strategy. In the words of one MTI leader, they struggled with the question, "How could we be that much out of touch with what was actually going on within our own society so that we did not play any role in society?"[48]

The response of Ghannoushi and other senior leaders in the wake of the labor unrest of January 1978 also underscored the tension within the Islamic movement between the older generation and its student followers, a tension over strategy and methods which would resurface at criti-

cal points in its history. While a communiqué was issued which supported the workers' demands, it also deplored the violence, a stance
which contrasted sharply with the more militant position of the movement's student supporters.[49]

Deteriorating conditions in Tunisia and the Islamic movement's position outside the political arena convinced Ghannoushi and the movement of the need to move beyond broad ideological statements and
relate Islam directly and specifically to the real, everyday political, economic, and social problems of the people. Islam must be seen as a source
not only of identity but also of true liberation—of the whole person and
society. This new direction placed the Islamists on a collision course with
the apolitical stance of the Quran Preservation Society: in 1978 they
were expelled, denounced as inflexible reactionaries.[50] The transformation of this relatively small group from a religiocultural force into a
sociopolitical movement was formalized with the establishment of the
Islamic Association (Jamaah al-Islamiyya) in 1979, one year after the
1978 riots and in the wake of Iran's revolution. Ideologically and organizationally, the Islamists were influenced by the example of the Egyptian
and Sudanese Muslim Brotherhoods. Thus a hitherto informal group of
like-minded people, run by its founders and key leaders and centered
around group meetings and discussions, became a more structured, participatory, activist organization with rules, guidelines, and a commitment to social action.

The Islamic Association was set up as a national organization guided
by its Amir (leader), Ghannoushi, and an elected consultative council
(*majlis al-shura*). Like many other Islamic activist organizations, the
organization had both a public and a secret private profile. The transformation of the Islamic movement into a sociopolitical organization directly involved in Tunisian political and economic affairs enhanced its
popularity but also invited government pressure. The realities of functioning under an increasingly hostile regime required that its cells be
secret and that much of its activities be underground. Ghannoushi
preached his message of a holistic Islam relevant to issues of political
and economic rights in the mosques of Tunisia. He correctly identified
the mind and pragmatic concerns of Tunisian youth:

> What concerns the young people of today? . . . The position of the
> Mutazilites [an early theological school or movement] on the attributes of
> God . . . whether the Quran is pre-existent or created? Was Islam re-
> vealed for this kind of useless, sterile argument? I wonder how our stu-
> dents feel studying "Islamic philosophy" when it offers them a bunch of
> dead issues having nothing to do with the problems of today. . . . I pro-
> pose that these shrouds be returned to their graves, that these false prob-

lems be buried and that we deal with our problems—economics, politics, sexual license.[51]

Tunisia's mosques hosted gatherings and provided platforms from which the struggle for Islam could be equated with the needs and grievances of the poor and oppressed. The movement also addressed these problems in communiqués, conferences, and the movement's publication *al-Marifa,* established in 1973, which provided religious legitimacy for the movement's sociopolitical message. Students and workers in particular were drawn to the new organization as either members or sympathizers. Ghannoushi and the Islamic Association spoke directly to current issues—workers' rights, jobs, wages, poverty, Westernization versus a more authentic national and cultural identity, political participation— presenting a living Islam rather than the "museum Islam" which Ghannoushi had encountered as a student.

Ghannoushi now combined his earlier role in France as a Muslim missionary and social activist among poor workers with that of a youth leader. The movement attracted most of its membership from the urban and rural working and lower middle classes. His message of political freedom and social activism in defense of the rights of the poor and the oppressed enabled the Islamic Association to align itself with the General Union of Tunisian Workers (UGTT) and win support within its ranks. At the same time it targeted universities and schools. Islamic student organizations grew and proliferated on campuses, attracting young men and women from the science and engineering faculties in particular. Although they clashed with secular nationalists and leftist students, Islamic student organizations eventually dominated student politics.

In April 1981, when the Bourguiba government briefly liberalized Tunisia's one-party political system, the Islamic Association was transformed into a political party, called the Islamic Tendency Movement (MTI), indicating that it was one of a number of trends (Islamic, secular, leftist, etc.) in Tunisian society. The phrase "Islamic Trend or Tendency" was already in use by Islamic university student groups, who had borrowed it from their Sudanese counterparts.

Although Bourguiba refused to issue a license legalizing the party, MTI continued to challenge the secular state. Its agenda included reassertion of Tunisia's Islamic–Arabic way of life and values, restriction of Tunisia's Westernized (Francophile) profile, and promotion of democracy, political pluralism, and economic and social justice. The brutality of Bourguiba's response to any and all opposition and the restrictive policies of his government alienated many Tunisians, spawning a religious and secular opposition which shared grievances and had both been equally repressed. MTI proved effective in attracting many of the disaffected, not only students, workers, and union members but also middle-class professionals, professors, teachers, engineers, lawyers, scientists, and doctors.

The politicization of the Islamic movement drew the wrath of Habib Bourguiba, who regarded it as a threat to his government. He had tolerated Islamic organizations as long as they remained apolitical, so as to counter the charges of his Islamic critics and the leftists. However, a political opposition was quite another matter. MTI's sociopolitical platform was a double challenge, for it echoed the criticisms of other (secular) opposition groups with whom it was willing to work and, even more threateningly, did so in the name of Islam. This challenge occurred at a time when much of the Muslim world was experiencing a resurgence of Islam and when many Muslim governments were nervous about the "Iranian threat"—Iran's declared intention to export its Islamic revolution.

In 1981 the Bourguiba government, which some experts had come to describe as a "modern administrative dictatorship,"[52] cracked down on MTI, arresting and imprisoning Ghannoushi and many of its leaders. Many were tortured in prison; those who escaped went underground or into self-imposed exile abroad. Despite the government's attempt to paint MTI as retrogressive, fundamentalist fanatics and a violent Iranian-backed revolutionary movement, Ghannoushi had distanced the movement from the excesses of the revolution and advocated a Tunisian rather than an Iranian solution. Ghannoushi denounced the use of violence and instead chose to work within the system, emphasizing a gradual process of social transformation and political participation as the means to realize MTI's long-range goal of establishing an Islamic state. He had been careful to talk about institutional reform rather than violent revolution, about change from below rather than the imposition of an Islamic system from above, about Islamic reform rather than implementation of medieval formulations of Islamic law.[53]

During this period more than ever before, MTI came to realize the limitations of the Muslim Brotherhood's ideology, which was conditioned by its Egyptian origins and experience. There was a growing conviction about the need to address the particular conditions of Tunisia by developing an ideology, program, and solutions more suited to the Tunisian experience.

The imprisonment of many of its leaders spawned new leadership and initiated a process of reevaluation. Students were recruited from the mosques and schools. They were invited (*dawa*) to Islam and were taught about it. At the same time that they were exposed to Western ideas in science and technology, these students of Tunisia's postindependence generation were required to read books on Islam and also on communism, in order to confront the communists. Looking for new solutions, many adapted well both to new Western scientific ideas and to an Islamic orientation. Thus they provided an alternative to the communists and the nationalists on campus.

Like Anwar Sadat and the Shah of Iran, Bourguiba refused to rein in his authoritarian tendencies. In later life he had exiled his wife, son, and prime minister. Moreover, an Islamic opposition proved doubly distasteful, representing for him an anachronistic retreat to the past which flew in the face of all that he had tried to represent, and a potentially potent popular force as well. Many rulers had become accustomed to branding their opposition Marxist as a smoke screen for their repression and to curry favor and support from the West. Bourguiba now joined those who in the eighties found a new global threat, the danger of "radical fundamentalism," in particular that of Iran, behind any challenge to their authority. However, "despite attempts to link the Islamist movement with Iranian extremism, [the Tunisian government] had not been able to eradicate what Bourghiba routinely and contemptuously described as the vestiges of an outdated religious traditionalism."[54] The image of Tunisia as a shining example of enlightened development, like that of Lebanon before it, seemed to crack if not shatter as the Bourguiba government attempted to stave off further popular strikes and riots and clashed with Islamic activists and professional syndicates.

Bread riots in 1984 forced Bourguiba to release Ghannoushi and other political prisoners after three years of imprisonment, as part of his attempt to ease tensions. However, government pressure on MTI continued.

> The 1984 amnesty did not end harassment. . . . The government banned civil servants from praying during work hours and closed mosques it had opened previously to buffer "leftist extremism." Public institutions were ordered not to hire back Islamists who had lost their jobs during the 1981–84 incarceration. Women wearing the veil were barred from the universities and workplaces. Islamist university students were expelled and drafted into the military. Taxi drivers caught wearing neatly trimmed beards—the mark par excellence of the Islamist—or listening to Islamist cassettes had their beards cut and their licenses revoked.[55]

Ghannoushi, although freed, was forbidden to speak in mosques or other public gathering places, nor was he allowed to teach, publish his writings, or travel abroad. In response to the prison experience of 1984–87, MTI was restructured with emphasis on its secret as well as public dimensions and a decentralized leadership. Along with a greater emphasis on clandestine cells, Ghannoushi's leadership role as the primary ideologue, though not the sole leader, was reinforced by the creation of a cadre of leaders across the country. Security measures which assured that no single individual would hold all the power or knowledge of MTI membership and activities were instituted.

Within three years after Ghannoushi's emergence from prison in 1984, Bourguiba once again moved against MTI. In March 1987 he arrested Ghannoushi and set off street battles and clashes between Islamic activist students and leftists in the universities. When the French government arrested six expatriate Tunisians for possession of arms, Bourguiba took the occasion to move once more against MTI, charging it with an Iranian-inspired plot to overthrow the government, although France had established no MTI connection and in fact claimed the six were members of the pro-Iranian Islamic Jihad and Hizbullah. More than three thousand MTI members were arrested. When Islamic Jihad subsequently claimed responsibility for a series of hotel bombings, ten members of MTI, not Islamic Jihad, were charged, despite the fact that the majority had already been imprisoned at the time of the alleged crime. Although the Bourguiba government tried to discredit MTI as "Khomeinists," Dirk Vanderwalle observed:

> The Islamic Tendency Movement . . . posed no threat to the country's rulers or its political system. But the movement's criticism of personal power, economic mismanagement, corruption, and moral laxity allowed it to become a symbol—perceived especially by the younger, educated generation as an alternative in a country void of political alternatives.[56]

Bourguiba was so intent upon eradicating his "Islamic threat" that when the courts sentenced Ghannoushi to life imprisonment at hard labor but not death, he ordered a new trial.

The 1981 and 1987 repression of MTI put severe strains upon the leadership and its moderate stance and exacerbated differences within the movement. As rumors spread that Bourguiba was intent upon Ghannoushi's execution and the eradication of MTI, MTI's members divided over the proper strategy; voices were increasingly raised advocating a tough response to government oppression. In November 1987, as the debate raged and a popular uprising seemed likely, Zeine Abedin Ben Ali, Bourguiba's prime minister, seized power from the aging dictator.

Ben Ali moved quickly to legitimate his rule and counter the government's Islamic opposition. He promised political liberalization, democratization, and a multiparty political system, precisely what all the opposition parties had been clamoring for. Moreover Ben Ali, reminiscent of

Egypt's Anwar Sadat, the Sudan's Gaafar Nimeiri, and Pakistan's Zulfikar Ali Bhutto, sought to enhance his legitimacy, broaden his base of popular support, diffuse criticism, and thus lessen the appeal of Islamic activists through an appeal to Tunisia's Arab–Islamic heritage. Ben Ali went on a much-publicized pilgrimage to Mecca; his speeches incorporated Islamic formulae; television and radio broadcast the call to prayer; the theological faculty at the Zaytouna mosque was reopened; the fast of Ramadan was officially observed; and Ben Ali promised to allow MTI to again publish its own newspaper or journal.

MTI responded to the promised political liberalization and movement toward democracy by offering to work with the new leader and participate in his call for a National Pact, in exchange for official recognition as a political party. Although Ben Ali, as Bourguiba's minister of internal affairs, had overseen the attempt to crush MTI, MTI's leadership took the risk. Most observers anticipated government recognition of the Renaissance Party, but in December 1989 Ben Ali categorically ruled out any political recognition for it, claiming that this decision "emanates from our firm belief in the need not to mix religion and politics, as experience has shown that anarchy emerges and the rule of law and institutions is undermined when such a mixing takes place."[57]

Two events in particular influenced Ben Ali's change of heart. The impressive performance of MTI candidates in national elections in 1988 demonstrated the party's political potential and validated its claim to be Tunisia's leading opposition group. This was followed by the Algerian government's movement toward political liberalization in the wake of riots and its recognition of Islamic political parties. The victory of the Islamic Salvation Front in the 1989 municipal elections realized the worst fears of Ben Ali and many Muslim rulers (and of many Western governments), who saw in political Islam not simply the threat of liberalization or democratization of their regimes but that of a "fundamentalist" victory.

Though the Renaissance Party (formerly MTI) avoided confrontational politics in the early years of Ben Ali's rule, the government's decision not to recognize it as a legal party, and its growing use of force and violation of human rights to intimidate its members, led the Renaissance Party to denounce the government for authoritarianism and the undue influence of the "secular left."[58] The confrontation between the government and the Renaissance Party escalated. Student demonstrations and strikes were crushed; Renaissance Party leaders were imprisoned or harassed. The deterioration of Tunisia's human rights record drew criticisms from human rights organizations. Ben Ali's political liberalization program proved far from democratic. As *The Economist* observed, "His party, the Constitutional Democratic Rally, rigged the poll in the 1989 general election and took every seat in parliament. Far from legalizing the leading opposition group, the Islamic party, Ennahda [Renaissance Party], the president has sought to crush it. . . . There is no real democracy and no press freedom."[59]

The Gulf War marked a turning point, as Ben Ali moved against the Renaissance Party in a "transparent effort to discredit it."[60] The war ushered in a period which would intensify the clash between the government and the Renaissance Party, a reputed coup attempt, and a split in its leadership with the defection of Abdelfattah Mourou. In May 1991, shortly after security forces had killed several students at a university demonstration, the Ben Ali government charged that it had uncovered a Renaissance Party plot, which included members of the Tunisian military, to seize power and establish a theocratic state. More than one third of those arrested were members of the Tunisian military reputed to be members of the Renaissance Party's secret military wing. Ben Ali called for unity in the face of a fundamentalist threat.[61]

Tunisia's Renaissance Party provides an example of the radicalization of movements in response to government manipulation of the political system, suppression, or violence. Increased government repression intimidates, factionalizes, and radicalizes. The result has been an escalation of confrontation and violence.

The varieties of Islamic activist groups and experiences are a testimony to the flexibility of Islam, and political Islam in particular. They illustrate the extent to which specific contexts, differences of political economy, Islam's capability of multiple and varied interpretations, and the distinctive personalities and ambitions of its leaders or ideologues shape the ideology and strategy of Islamic movements. This diversity underscores the problem of terminology, the extent to which facile labels can become an obstacle to understanding. How does one encompass in one term (revivalism, fundamentalism, militant Islam) the meanings and usages of Islam by rulers (Qaddafi, Khomeini, Zia ul-Haq, Nimeiri) and Islamic organizations, their differing attitudes toward and relationships with the West, as well as the diversity of strategies and goals employed by mainstream activists versus violent radical revolutionaries? Islamic politics must be viewed within specific country contexts; far from being a monolithic reality, it manifests a rich diversity of leaders and forms.

A second obstacle in assessing the future impact of Islamic movements—including the extent to which they will be a challenge or a threat—is their lack of a well-developed political agenda. Theoretical or ideological statements are often not accompanied by specific models for change. Islamic movements tend to be more specific about what they are against than what they are for. While all may speak of an Islamic order or state, of implementation of the Sharia, of a society grounded more firmly on Islamic values, the details are often vague. This has been true for older movements such as Egypt's Muslim Brotherhood as well as for more recent groups. After the electoral victory of Algeria's FIS (Islamic Salvation Front), when Abbasi Madani was questioned about his program, he described it as broad. When pressed for its practical steps, he replied, "Our practical program is also broad." Similar experiences are true for other Islamic movements. Though many speak of Islamic government, they differ in their understanding of its nature, speaking about a variety of forms—a caliphate, consultative government and a strong executive, or a multiparty system— but do not go beyond this. Most activist leaders admit that, should they gain power, they would lack personnel, paradigms or models (political, economic, legal), and experience. Realizing their current unpreparedness to govern, some are more comfortable in the opposition or in parliament. Their critics—Muslim governments and Western policymakers—often seize upon this failure to be specific. Ironically, many critics argue that the extent to which movements represent only a minority of the population and have no real program dooms them to failure. Yet none seems willing to provide them with the opportunity to discredit themselves. Instead, despite fits of democratic rhetoric, political liberalization and democratization have been placed on hold or subverted, while Western governments are generally content to look the other way.

At the micro level, Islamic organizations have been successful in responding concretely to people's problems. Their Islamic ideologies offer a framework for meaning and purpose that responds to issues of identity, faith, and authenticity. As social movements they respond to the needs of many by providing educational and social welfare services. Islamic movements may at times be antiforeign, but seldom antimodern. Social gains have been accompanied by successes in the political arena as well. The Islamic movement has bridged the mythical divide between religion and secular modernization in the cities and towns of the more advanced countries of the Muslim world: Iran, Egypt, Lebanon, and Tunisia.

Islamic organizations often enjoy an impact disproportionate to their actual size and numbers. They have features that few political parties are able to match. They are well organized, disciplined, and highly motivated. While some advocate violence and engage in terrorism, many today, if given the opportunity, participate within the political system. Many exemplify a high degree of integrity and self-sacrifice, have a clear sense of purpose and commitment, foster a sense of certitude and God-ordained mission. Activist organizations prove especially attractive because they offer an indigenously based and religiously legitimated alternative. They exude a strong sense of self-confidence and identity to those who are dissatisfied and disillusioned, who experience their governments and society as inadequate, corrupt, and incapable of responding to their people's political and socioeconomic needs. Islamic organizations and parties have proven especially successful where failed economies, unemployed and disaffected youth, and repressive governments make them effective critics of regimes. Their voices for political liberalization and social change often represent the only credible alternative to the government. The electoral record of Islamic organizations and their emergence as the major opposition in Egypt, Algeria, Tunisia, and Jordan illustrate this reality.

Despite stereotypes of activists as fanatics who wish to retreat to the past, the vast majority share a common call for the transformation of society not through a blind return to seventh-century Medina but a response to the present. They do not seek to reproduce the past but to reconstruct society through a process of Islamic reform in which the principles of Islam are applied to contemporary needs. Each speaks of a comprehensive reformation or revolution, the creation of an Islamic order and state, since they regard Islam as comprehensive in scope, a faith-informed way of life.

The modern nation-state is seen as in collapse; modern society is often described as in disorder (*fitna*). Why? Both the West and the traditional religious establishment (coopted preachers of a static Islam) are denounced as the culprits. They especially blame Western models of development and Westernized ruling elites—what Rashid Ghannoushi has called the "dictatorship of the ruling elite minority." A tyranny of the few over the many has led to political and economic dependence on the West and loss of cultural identity. All activist movements advocate a process of resocialization or re-Islamization of the individual (greater religious education and awareness) and mobilization of the community's political, economic, and administrative systems. The process is one of *dawa* and jihad, the call to be more self-consciously Islamic and the struggle to implement Islam in personal life and in state and society.

Despite their critique of the nation-state, most Islamic movements today, like most Muslims, accept its reality, but within the broader context of Islamic universalism they look toward the *ummah*. Like early Islamic modernists such as Abduh and Iqbal, they believe national reform is not contrary to, but a necessary step in and part of, the reconstitution of Islamic universalism. All advocate independence from the West and the East politically, economically, and culturally. However, they vary in their posture toward the West. Some, such as Hizbullah and many of the followers of the Ayatollah Khomeini, have totally rejected the West. Others like Egypt's Muslim Brotherhood, AMAL, and the Renaissance Party do not preclude relations with or selective borrowing from the West.

Most movements have increasingly taken great care to identify themselves with the disinherited or oppressed (*mustadhafin*) and emphasize issues of social justice. The socioeconomic situations in their own countries as well as the Iranian example have been formative influences. The plight of the poor, the condemnation of corruption, disparities of class and wealth, and the failure of the economy in general, whether it be capitalism or state socialism, are important to their populist messages.

In recent years most Islamic movements have moved toward a populist, participatory, pluralistic political stance, championing democratization, human rights, and economic reform. While many were critical of "democracy" in the past, they claim that this was a response to the secularism of the modern state as well as the autocratic nature of Muslim regimes. To varying degrees in recent years, most Islamic movements have emphasized change not through violence (which many now view as counterproductive) but through the political and social transformation of society. They speak of the need to prepare people for an Islamic order rather than to impose it. They seek recognition of political rights and participation in the electoral process. Yet there are differences within and among different groups. In North Africa, for example, Morocco's Abdessalem Yassine and Algeria's Abbasi Madani have been slower than Ghannoushi to recognize multiparty politics and the democratic process. Moreover, the history of Islamic movements in Egypt, Algeria, and Tunisia has shown that, where regimes deny political participation or suppress movements, state violence often begets violence. As the recent examples of the Renaissance Party and FIS demonstrate, there are diverse tendencies within Islamic movements; state repression and violence—suppression of demonstrations, denial of free and fair elections, arrests, imprisonment, torture—exacerbate differences, threaten to divide organizations, and often result in violent responses in what is regarded as legitimate self-defense against state oppression.

Just as the political economies and national experiences (from policies of cooptation and accommodation to violence and repression) of Muslim countries vary, determining the differing relations between governments and Islamic organizations, so too differences in ideological orientation and strategy exist within Islamic movements. All major Islamic leaders or ideologues and movements have emphasized and been influenced by belief in the need to construct their own solutions to their specific contexts and local problems. There are many such organizations; equally important, within each movement—the Renaissance Party and FIS, for example—there are different currents of thought. Madani's comment, "There is diversity in FIS but it is united," has been echoed by Ghannoushi. This diversity extends to attitudes toward modernization, militancy, the West, democratization and pluralism, the role of women, and most recently, the Gulf War and the use of force in domestic politics. Just as MTI experienced the defection of the Progressivists in the past, so too more recently it has been faced with a new split. Ghannoushi and many other party members differed regarding support for Iraq during the Galf War. Most recently, Ghannoushi and some other Renaissance Party leaders, in particular Abdelfattah Mourou, have split over tactics (in particular the use of violence) in responding to government repression. In Algeria's FIS, Madani and Ali Benhadj have differed sharply in their public statements about democracy. Moreover, Madani's statement—"There is no other way at present . . . the way to power is elections which are decided through the popular will of the people"[62]—does not answer those who wonder whether the espousal of democracy by Islamic movements is a pragmatic, tactical accommodation or a principled position. Of course the same could be said as well for many incumbent Muslim governments and secular elites that advocate political liberalization and democracy.

6

"Islamic Fundamentalism" and the West

The Muslims are coming, the Muslims are coming! A caricature of Western fears?[1] Exaggerated? Perhaps. However, when Dan Quayle, the vice president of the United States, speaks of the danger of radical Islamic fundamentalism, grouping it with Nazism and communism, and magazines and newspaper editorials speak of Islam's war with the West and its incompatibility with democracy, and a respected national newspaper, the *Boston Globe,* runs a four-part series on Islam whose general tenor is captured by the title of its introductory piece, "The Sword of Islam," it is difficult to know where reality ends and myth begins.[2]

There are lessons to be learned from our Soviet experience. Celebration of the unraveling of communism and the victory of democracy has been tempered by questions that go to the heart of our ability to understand, analyze, and formulate policy. Delight at the triumph of democracy has been accompanied by a growing realization of the extent to which fear and the demonizing of the enemy blinded many to the true extent of the Soviet threat. Viewing the Soviet Union through the prism of the "evil empire" often proved ideologically reassuring and emotionally satisfying, justifying the expenditure of enormous resources and the support of a vast military-industrial complex. However, our easy stereotypes of the enemy and the monolithic nature of the communist threat also proved costly. Despite an enormous amount of intelligence and analysis, few seemed to know that in the end the emperor had no clothes. Neither government agencies nor academic think tanks predicted the extent and rapidity of the disintegration of the Soviet empire. The exaggerated fears and static vision which drove us to take herculean steps against a monolithic enemy blinded us to the diversity within the Soviet Union and the profound changes that were taking place.

Similarly, in understanding and responding to present-day events in the Muslim world, we are again challenged to resist easy stereotypes and solutions. There is an easy path and a hard one. The easy path is to view Islam and Islamic revivalism as a threat—to posit a global Pan-Islamic threat, monolithic in nature, a historic enemy whose faith and agenda are diametrically opposed to the West. This attitude leads to support for secular regimes at almost any cost rather than risking an Islamically oriented government in power.

The more difficult path is to move beyond facile stereotypes and ready-made images and answers. Just as simply perceiving the Soviet Union and Eastern Europe through the prism of the "evil empire" had its costs, so too the tendency of American administrations and the media to equate Islam and Islamic activism with Qaddafi/Khomeini and thus with radicalism, terrorism, and anti-Americanism has seriously hampered our understanding and conditioned our responses. As Patrick Bannerman, a former diplomat and analyst for the British Foreign Office, observed in his book *Islam in Perspective,* "How non-Muslims think of Islam conditions the manner in which they deal with Muslims, which in turn conditions how Muslims think of and deal with non-Muslims."[3] The challenge today is to appreciate the diversity of Islamic actors and movements, to ascertain the reasons behind confrontations and conflicts, and thus to react to specific events and situations with informed, reasoned responses rather than predetermined presumptions and reactions.

In the post–Gulf War period, the struggle to articulate a New World Order has collided with an inability to move beyond the stereotypes and biases of the past. Exhilaration over the prospect of democratization in Eastern Europe contrasts sharply with the fear or at best silence of the U.S. government and indeed many in the West at the prospect of similar demands by Islamic populist movements in the Arab and broader Muslim world. The failure to speak out forcefully for democratization in the Middle East and to condemn government repression against Islamic movements in Tunisia, Algeria, and Egypt in effect denies Islamic activists the political participation and human rights enjoyed by others and raises serious questions about a "selective" approach to democratization. At the heart of this approach is a tendency to define stability in terms of support for the status quo in Muslim countries, however much it may contradict democratic values.

The easy excuse for looking the other way and continuing to support autocratic Middle Eastern regimes is to question whether Islam or Arab culture are indeed conducive to democratization. Ironically, similar concerns were not raised with regard to the Soviet Union and Eastern Europe. Was it because the peoples of Eastern Europe and the Soviet

Union, steeped in decades of communist rule and socialization, are somehow less prone to authoritarianism? Or was it because we believe that the common Judaeo-Christian cultural tradition of the West provided a more accommodating ideological soil, was more inherently democratic? Are we still inhibited by the baggage of the past? A combination of ignorance, stereotyping, history, and experience, as well as religio-cultural chauvinism, too often blind even the best-intentioned when dealing with the Arab and Muslim world. Of course similar comments could be made about the approach of many Muslims to the West. The result is a dual tendency to regard each other as a threat. Thus the theme of an impending confrontation can be found among many in the Muslim world and in the West.

As we have seen, though Islam is the second largest of the world's religions and the Judaeo-Christian tradition has strong historic and theological links with Islam, the history of Muslim–Jewish–Christian relations has been more one of competition and combat than dialogue and mutual understanding, driven by competing theological claims and political interests. The confrontations and conflicts have spanned the ages and reinforced images of a historic and global militant Islam: the early Muslim expansion and conquests; the Crusades and the fall of Jerusalem; Ottoman hegemony over Eastern Europe and, with the siege of Vienna, its threat to overrun the West; the great jihads against European colonial rule; Arab–Israeli wars; the economic threat of oil embargoes; Iran's humiliation of an "America held hostage" and its threat to export its revolution; media images of despots (Qaddafi, Khomeini, Saddam Hussein) wielding an Islamic sword and calling upon the frenzied faithful to rise up against the West; and the specter of radical revolutionary groups seizing Western hostages, hijacking planes, and wantonly visiting a reign of terrorism. Death threats against Salman Rushdie and Muslim secular intellectuals like the Egyptian philosopher Fouad Zakaria, who had declared that "the tide of political Islam . . . constitutes a very real danger," reinforce images of an intolerant and dangerous Islam.[4]

Saddam Hussein's call upon the world's Muslims to rise up and wage holy war against Western Crusaders was a chilling reminder to many of the Ayatollah Khomeini's threat to export Iran's Islamic revolution. It also confirmed fears of a militant, confrontational Islamic threat or war against the West. The support that Saddam enjoyed from leaders of Islamic movements in Algeria, Tunisia, the Sudan, and Pakistan reinforced the arguments of those who view Islam and the Muslim world as being on a collision path with Western priorities and interests.

Despite predictions by some experts that the political phenomenon identified as "Islamic fundamentalism" was a spent force, we see the continued vitality of an Islamic revivalism that extends from North Africa to Southeast Asia, in the streets and at the polling booths. This vision can reinforce the equation of the political challenge of Islam with a threat.[5] In the aftermath of the triumph of the Islamic Salvation Front in Algeria's municipal elections (and its subsequent sweep of national parliamentary elections), similar fears surfaced in other North African countries and in Europe.

> Across North Africa and especially in Tunisia and Morocco, leaders are anxiously waiting for further shock waves in the politicization of Islam, where banned fundamentalist movements threaten the status quo. There are similar fears in France, Belgium, Italy and Spain which have over 2 million Algerians. Meanwhile the victory of the fundamentalists is alienating foreign investors, who equate Islam with political and economic risk.[6]

The West often reinforces the equation of Islam with danger or threat, viewing the Islamic world with catchwords like "militant Islam," "Islamic fundamentalism," and "terrorism." Our selective memory then blocks our ability to appreciate the other side of the equation—the sources of Muslim images of the West in turn as the "real" threat to them. Many in the Arab and Muslim world view the history of Islam and of the Muslim world's dealings with the West as one of victimization and oppression at the hands of an expansive imperial power. Thus many counter that it is "militant Christianity" and "militant Judaism" that are the root causes of failed Muslim societies and instability: the aggression and intolerance of Christian-initiated Crusades and the Inquisition; European colonialism; the breakup of the Ottoman empire and artificial creation of modern states in Iraq, Lebanon, Syria, Transjordan, and Palestine; the establishment of Israel; Israel's occupation of the West Bank and Gaza and its invasion and occupation of Lebanon; and the extent to which oil interests have been the determining factor in support for autocratic regimes.

The realities of colonialism and imperialism, although forgotten or conveniently overlooked by many in the West, are part of its living legacy, firmly implanted in the memory, however exaggerated at times, of many in the Middle East. As the Iranian Revolution demonstrated, several decades had not erased the memories and humiliation of imperial intervention: the willingness of the Soviet Union to invade a "neutral" Iran in 1941 and, shortly thereafter, the British and Soviets forcing Reza Shah Pahlavi to abdicate in favor of his son, Muhammad Reza Shah Pahlavi, and finally U.S. intervention in Iranian politics in the early fifties when it returned the Shah from exile in Italy to Teheran. As we have seen, in recent years the memories of centuries of Western hegemony followed by continued dependence upon the West have left deep resentments which become easy excuses for societal failures and have proven combustible materials in Muslim politics. If there is an Islamic threat, there has also been a Western threat—of political and religiocultural imperialism, a political occupation accompanied by cultural invasion. As a result, many in the Muslim world, like their counterparts in the West, opt for easy anti-imperialist slogans and demonization. At its worst, both sides have engaged in a process of "mutual satanization."[7]

The Islamic Threat Today

Fear of Islam is not new. The tendency to judge the actions of Muslims in splendid isolation, to generalize from the actions of the few to the many, to disregard similar excesses committed in the name of other religions and ideologies (including freedom and democracy), is also not new. Maxime Rodinson has compared the historic polarized relationship between Christianity and Islam to the more recent worldwide competition of capitalism and communism:

> From a political and an ideological perspective, if one compares the attitudes of Christianity towards Islam with those of Western capitalism and communism today, the parallels are clear. In each grouping, two systems are at odds: yet within each system, a single dominant ideology unites divisive and hostile factions.[8]

In some ways, the attitude of the West toward communism seems at times transferred to or replicated in the new threat, "Islamic fundamentalism." Indeed, in the nineties the effects of this polarization are expressed by the prevailing tendency of governments in the Muslim world and the West, the media, and many analysts to conclude, without regard to the diversity of Islamic organizations and specific social contexts, that Islamic fundamentalism is inherently a major global threat.

Cognizant of a Western tendency to see Islam as a threat, many Muslim governments use the danger of Islamic radicalism as an excuse for their own control or suppression of Islamic movements. They fan the fears of a monolithic Islamic radicalism both at home and in the West, much as many in the past used anticommunism as an excuse for authoritarian rule and to win the support of Western powers. The banning of

Islamic organizations, imprisonment of activists, and violation of human rights are excused with the plaintive plea, "We are facing young fanatics threatening our future."⁹ Western stereotypes of a united worldwide fundamentalist movement that threatens the stability of the Arab world and Western interests are exploited by Arab diplomats from states with strong Western ties who declare: "Fundamentalism is international in scope. It has branches everywhere. . . . Fundamentalist expansion will eventually threaten the industrial nations when most Arab countries have been destabilized."¹⁰ The focus on radicalism and the equation of Islam with an extremism that threatens to confront the West has become commonplace. Too often we are exposed in the media and literature to a sensationalized, monolithic approach which reinforces facile generalizations and stereotypes rather than challenging our understanding of the "who" and the "why" of history, the specific causes or reasons behind the headlines. In government and professional discussions, moreover, as was the case with McCarthy-era anticommunism, not to be simply dismissive of Islamic activism is often viewed as being biased or sympathetic toward the enemy.

A "selective" presentation and analysis of Islam and events in the Muslim world by prominent scholars and political commentators too often inform articles and editorials on the Muslim world. This selective analysis fails to tell the whole story, to provide the full context for Muslim attitudes, events, and actions, or fails to account for the diversity of Muslim practice. While it sheds some light, it is a partial light that obscures or distorts the full picture. As a result, Islam and Islamic revivalism are easily reduced to stereotypes of Islam against the West, Islam's war with modernity, or Muslim rage, extremism, fanaticism, terrorism. The "f" and "t" words, "fundamentalism" and "terrorism," have become linked in the minds of many. Selective and therefore biased analysis adds to our ignorance rather than our knowledge, narrows our perspective rather than broadening our understanding, reinforces the problem rather than opening the way to new solutions.

The stereotypical image of Islam and Muslims as menacing militant fundamentalists was reflected strikingly in Prof. Bernard Lewis's talk entitled "Islamic Fundamentalism," given as the prestigious Jefferson Lecture of 1990, the highest honor accorded by the U.S. government to a scholar for achievement in the Humanities. A revised version became a lead article, "The Roots of Muslim Rage," in the *Atlantic Monthly.* The "packaging"—the new title, the magazine's cover, and the article's two pictures—of "Roots of Muslim Rage" reflects the pitfalls of a selective presentation. It reinforces stereotypes of Islamic revivalism and of

Muslims and predisposes the reader to view the relationship of Islam to the West in terms of rage, violence, hatred, and irrationality. Because of Bernard Lewis's international stature as a leading scholar and political commentator on the Middle East, his topic, and its prominent public platform, "Roots of Muslim Rage" received widespread coverage nationally and internationally. It has had a significant impact both on Western perceptions of contemporary Islam and on many Muslim perceptions of how Islam and Muslims are viewed in the West.

The message and impact of "Roots of Muslim Rage" is reinforced by the picture on the front cover of the *Atlantic Monthly,* portraying a scowling, bearded, turbaned Muslim with American flags in his glaring eyes. The threat motif and confrontational tone are supplemented by the two pictures used in the article, ostensibly presenting the quintessential Muslim perception of America as the enemy. The first is of a serpent marked with the stars and stripes seen crossing a desert (America's dominance of or threat to the Arab world); the second shows the serpent poised as if to attack from behind an unsuspecting pious Muslim at prayer.[11] Like other sensationalist stereotypes, pictures meant to be provocative, to attract the reader, feed into our ignorance and reinforce a myopic vision of the reality. Muslims are attired in "traditional" dress, bearded and turbaned, despite the fact that most Muslims (and most "fundamentalists") do not dress or look like this. The result reinforces the image of Islamic activists as medieval in life-style and mentality.

The title, "Roots of Muslim Rage," sets the tone and expectation. Yet would we tolerate similar generalizations in analyzing and explaining Western activities and motives? How often do we see articles that speak of Christian rage or Jewish rage? In a similar vein, the nuclear capability of Muslim countries such as Pakistan has often been spoken of in terms of an "Islamic bomb," implying the existence of a monolithic Muslim world threatening Israel and the West. Do we expect Israel's or America's nuclear capabilities to be described in terms of a Jewish or a Christian bomb? Some Muslims have described Israeli bombings of Beirut as the result of "Jewish boys dropping Christian bombs"—a description which most in the West would find inaccurate and offensive.

There is a lesson to be learned from the failure of talented analysts who continued to warn of the dangers of a monolithic communist threat while the Soviet Union was in fact an economic basket case, breaking apart from within. Partial analysis which reinforces comfortable stereotypes and Western secular presuppositions must be transcended, if we are to avoid the ideological pitfalls and biases of a political analysis driven by an exaggerated threat.

Islam Against the West, Islam in the West

For a millennium, the struggle for mankind's destiny was between Christianity and Islam; in the 21st century, it may be so again. For, as the Shiites humiliate us, their co-religionists are filling up the countries of the West.

—PATRICK BUCHANAN

According to many Western commentators, Islam and the West are on a collision course. Islam is a triple threat: political, demographic, and socioreligious. For some, the nature of the Islamic threat is intensified by the linkage of the political and the demographic. Thus Patrick Buchanan could write that while the West finds itself "negotiating for hostages with Shiite radicals who hate and detest us," their Muslim brothers are populating Western countries. The Muslim threat is global in nature as Muslims in Europe, the Soviet Union, and America "proliferate and prosper."[12] Other observers such as Charles Krauthammer, in the midst of the unraveling of the Soviet Union, spoke of a global Islamic uprising, a vision of Muslims in the heartland and on the periphery of the Muslim world rising up in revolt: a "new 'arc of crisis' . . . another great movement is going on as well, unnoticed but just as portentous: a global intifada."[13]

Much as observers in the past retreated to polemics and stereotypes of Arabs, Turks, or Muslims rather than addressing the specific causes of conflict and confrontation, today we are witnessing the perpetuation or creation of a new myth. The impending confrontation between Islam and the West is presented as part of a historical pattern of Muslim belligerency and aggression. Past images of a Christian West turning back the threat of Muslim armies are conjured up and linked to current realities: Charles Martel, who "halted the first Mohammedan advance, preventing the crescent from closing over Christian Europe," the Crusaders' attempt to save Jerusalem, and the narrow defeat of "Islamic legions" at Vienna, are linked to current realities and the proclamation, "Now, Islam is again resurgent." A combination of radicalism and spiraling population growth threatens to overrun East and West: "Clearly, Islam is in the ascent in Africa, Asia and the Middle East. In the West, devout Moslems are having children, while in our secular societies, the philosophy of Planned Parenthood takes hold and the condom is king."[14]

Events in the West continue to reinforce the perspective of an impending demographic threat. As Europe experiences a new wave of immigration which now supplements earlier migrant worker communities, settler workers (from Romania, Albania, North Africa, Turkey) have become an explosive political issue. The impact and assimilation of Muslims is a particularly contentious issue in countries such as Great Britain and France, where earlier settled communities from the Middle East, Asia, and Africa have become sizable, dominating local politics in cities like Bradford and Marseilles, and asserting their rights nationally and locally. These communities now threaten to grow significantly, supplemented by new waves of immigrants. The presence of significant Muslim minority populations puts strains on the social fabric of European societies such as France, where Islam is the second largest religion, and Great Britain, where it ranks third. Anti-Arab/Muslim sentiment in Western Europe is part of a growing xenophobia. Muslim communities and indigenous groups have clashed over questions of continued immigration, citizenship, and the accommodation of Muslim belief and practice. In England debates ranging from the Salman Rushdie affair to local politics and school issues have raged. Muslims have demanded that Muslim schools be allowed to receive a state subvention, as is allowed for Catholic and Jewish schools. The government has repeatedly refused to do this, usually claiming that these schools were "substandard." Such events have contributed to British Muslim politicization as well as evidenced the diversity of Muslim responses. While some British Muslims have created a Muslim Parliament of Great Britain, which includes a Speaker and four deputy speakers (including two women), there has been considerable dissension within Muslim ranks in Britain: some have charged that this is merely playing the Western game (i.e., "Parliament"); others, that it constitutes a form of "apartheid."[15] Still other Muslim leaders have objected to any notion of separatism as an alternative to their integration into British society.

> Many Muslims would agree with the grievances that Dr. Siddiqui [Kalim Siddique, the architect of the Muslim Parliament of Great Britain] identifies—for instance, racism, an unequal share of unemployment, etc.—but the idea of separatism is anathema to most Muslims. Their real grievance is that they do not find it easy to get into the mainstream of British society.[16]

In France, calls for the expulsion of foreign workers have been accompanied by a celebrated case in which Muslim girls were prohibited from wearing a head scarf in school. Ironically, political demagogues in France play on many of the same fears that are so prevalent in the Muslim world: "Europeans now feel threatened by so many things. . . . The Americanization of our culture, the drift to the left, a loss of identity, the rise of an ugly populist response to things foreign."[17]

Western European fears often express themselves in religiocultural as well as economic and political terms. While assimilation of other Europeans occurred in the past, many today doubt the possibility or desirability of absorbing the new immigrants, especially Muslims who are both nationally and religioculturally alien: "Our former immigrants were Europeans; these are not. Arab girls who insist on wearing chuddars [the chador, veil or covering] in our schools are not French and don't want to be. . . . Europe's past was white and Judaeo-Christian. The future is not. I doubt that our very old institutions and structures will be able to stand the pressure."[18] Some observers see the outcome of the current debate in potentially cataclysmic terms: "While Europe has overcome the cold war . . . it now risks creating new divisions and conflicts, such as a white, wealthy and Christian 'Fortress Europe' pitted against a largely poor, Islamic world. That could lead to terrorism and another forty years of small, hot wars."[19]

The Muslim presence in Europe and North America does pose a challenge, though not in the sense that Buchanan and others have predicted. Whether or not they will adapt to our liberal agenda is indeed an important question and cannot be dismissed. But one can also ask: will we adapt our own liberalism? Liberalism has long accepted that ethnicity and race are so inherent to the human condition that discrimination along these lines is unacceptable. Yet with our secularized mentality, we have failed to appreciate that, for many people in the world, religious faith is also a primary identity. It is a given, not a choice, and as such Muslim citizens in the West cannot be expected to forgo certain rights in society (e.g., school accommodation to dietary laws, dress codes, and holy days, and the right of workers to observe their "sabbath" by attending the Friday congregational prayer). Thus when we ask what kind of democrats they are, we must be prepared also to ask and answer, what kind of democrats are we? To some extent it is now true that, increasingly in Western societies, "Islam is us."[20]

Patrick Buchanan's vision of an impending conflict between Islam and the West as but another stage in a historic (if not inevitable) pattern of confrontation is reinforced by Bernard Lewis's "Roots of Muslim Rage." As the *Atlantic Monthly* noted, paraphrasing Lewis, in its table of contents:

> The struggle between Islam and the West has now lasted fourteen centuries. It has consisted of [rather than included] a long series of attacks and counterattacks, jihads and crusades, conquests and reconquests. Today much of the Muslim world is again seized by an intense—and violent— resentment of the West. "Suddenly," a distinguished historian [Lewis] of Islam writes, "America had become the archenemy, the incarnation of

evil, the diabolic opponent of all that is good, and specifically, for Muslims, of Islam. Why?"[21]

Islam and Muslims are portrayed as the instigators and protagonists in
fourteen centuries of warfare. Islam is the aggressor. Thus in the above
statement, Islam and the acts of Muslims are described as aggressive—
responsible for attacks, jihads, and conquests—while the West is described as defensive, responding with counterattacks, crusades, and re-
conquests. Despite the portrayal of fourteen continuous centuries of
confrontation, the reader is informed that "suddenly" America has become the archenemy, evil personified, and so forth. If the contemporary
threat is "sudden," then the reader will logically conclude that Muslims
have a historic propensity to violence against and hatred of the West, or
else that Muslims are an emotional, irrational, and war-prone people.
    Throughout most of the discussion of the "Islamic threat" or the roots
of Muslim rage, there is a surprising paucity of information and lack of
discrimination concerning the nature and diversity of the Islamic resurgence. Professor Lewis in "Roots of Muslim Rage" tells us what the roots
of Muslim rage are, but very little about who these Muslims are. There is
little recognition that Islamic organizations or activists, while sharing a
general ideological commitment, do in fact differ among themselves.
References in his article to specific organizations are minimal, as is any
discussion of the education, social backgrounds, and activities (other than
violence and terrorism) of their members. There is little to contradict the
stereotype of strange, backward-looking figures from another age. The
reader never learns why Islamic activism proves an attractive option for
many educated Muslims. The predominant picture presented is that of
radicalized, marginalized, and often violent revolutionaries, traditional
in dress and at war with modernity. The reader never learns that the
majority of so-called fundamentalist organizations are urban-based and
led by well-educated leaders who attract students and followers and are
well placed in the professions (engineering, science, medicine, law, education, the military), or that they provide social and medical services and
have functioned effectively within the political system.
    The selective representation of most analyses of Islamic activism
omits, downplays, or dismisses the reasons given by activists (and indeed
by many Arabs and Muslims) for their criticism and rejection of the
West: imperialism, America's tilt toward Israel, and its support for oppressive regimes (the Shah's Iran, Nimeiri's Sudan, Lebanon). The
reader is never challenged to consider the reasons behind the attitudes
and actions of activists, rather than simply dismissing them as provoked
by a clash of civilizations or a blind, irrational clinging to faith.

Throughout most of the discussion of the "Islamic threat" or the roots of Muslim rage, there is a surprising paucity of information and lack of discrimination concerning the nature and diversity of the Islamic resurgence. Professor Lewis in "Roots of Muslim Rage" tells us what the roots of Muslim rage are, but very little about who these Muslims are. There is little recognition that Islamic organizations or activists, while sharing a general ideological commitment, do in fact differ among themselves. References in his article to specific organizations are minimal, as is any discussion of the education, social backgrounds, and activities (other than violence and terrorism) of their members. There is little to contradict the stereotype of strange, backward-looking figures from another age. The reader never learns why Islamic activism proves an attractive option for many educated Muslims. The predominant picture presented is that of radicalized, marginalized, and often violent revolutionaries, traditional in dress and at war with modernity. The reader never learns that the majority of so-called fundamentalist organizations are urban-based and led by well-educated leaders who attract students and followers and are well placed in the professions (engineering, science, medicine, law, education, the military), or that they provide social and medical services and have functioned effectively within the political system.

The selective representation of most analyses of Islamic activism omits, downplays, or dismisses the reasons given by activists (and indeed by many Arabs and Muslims) for their criticism and rejection of the West: imperialism, America's tilt toward Israel, and its support for oppressive regimes (the Shah's Iran, Nimeiri's Sudan, Lebanon). The reader is never challenged to consider the reasons behind the attitudes and actions of activists, rather than simply dismissing them as provoked by a clash of civilizations or a blind, irrational clinging to faith.

The shift in Muslim attitudes toward the West from admiration and emulation to hostility and rejection is reduced by Bernard Lewis to a clash of separate and distinct (almost mutually exclusive) civilizations: "Fundamentalist leaders are not mistaken in seeing in Western civilization the greatest challenge to the way of life that they wish to retain or restore for their people."[22] Fundamentalists wage war against modernity: secularism, Western capitalism, democracy. The stereotypical dichotomies are reinforced: Islam against the West, fundamentalism against modernity, static tradition versus dynamic change, the desire to return to or preserve the past versus adaptation to modern life. All so-called fundamentalists are lumped together, obscuring the fact that many have a modern education, hold responsible professional positions, and participate in the democratic process.

Emphasis upon a clash of civilizations reinforces the tendency to downplay or overlook specific political and socioeconomic causes for Muslim behavior, to see Muslim actions as an irrational reaction rather than a response to specific policies and actions.

> It should now be clear that we are facing a mood and a movement far transcending the level of issues and policies and the governments that pursue them. This is no less than a clash of civilizations—the perhaps irrational but surely historic reaction of an ancient rival against our Judaeo-Christian heritage, our secular present, and the worldwide expansion of both.[23]

The primacy of competing political interests, policies, and issues is dismissed or eclipsed by the vision of an age-old rivalry between "them and us." Once again, history witnesses Islam pitted against the West, against "our Judaeo-Christian" and "secular" West, a confrontation of competing and conflicting global visions and missions. This approach is as valid and useful as the attempts of Saddam Hussein (like the Ayatollah Khomeini before him) to reduce the West's antagonism to him to Crusader imperialism against the oppressed masses of the Arab and Muslim worlds.

Denial of the need to address specific political or socioeconomic issues also negates any notion of shared responsibility. How can the West respond to what is obviously emotional and "irrational," an assault upon it by peoples who are peculiarly driven by their passions and hatred? In explaining this phenomenon, James Piscatori has observed:

> Whether it was the Ottoman attempt to thwart Christian nationalists or the Muslim attempt to gain independence from the West, Islam was fanatical because it ran counter to imperial interests. But it was the converse formulation that became the standard explanation of Muslim conduct:

> Islam was hostile to the West because it was fanatical. . . . Consequently,
> Muslims came to be seen as a uniformly emotional and sometimes illogical
> race that moved as one body and spoke with one voice.[24]

The pattern persists. As many in the past dismissed the political realities of Muslim–Christian relations by retreating to the bogey of "Muslim
fanaticism," so Muslims today are described as a people who, though
courteous and possessed of a rich religious tradition and civilization,
harbor a propensity to hatred and rage when angered:

> And yet, in moments of upheaval and disruption, when the deeper pas-
> sions are stirred, this dignity and courtesy toward others can give way to
> an explosive mixture of rage and hatred which impels even the govern-
> ment of an ancient and civilized country [Iran]—even the spokesman
> [Ayatollah Khomeini] of a great spiritual and ethical religion—to espouse
> kidnapping and assassination, and try to find, in the life of their Prophet,
> approval and indeed precedent for such actions.[25]

Many people—believers and nonbelievers, Muslims and Jews, Christians, Sikhs, and Hindus—become enraged when their survival or interests are threatened. Rage and hatred, or the use of religion to rationalize
and legitimate one's actions, are not peculiar to Muslims alone. In reality most "civilized" peoples, normally courteous and kind, accept rage
against evil and hatred of enemies as a normal response to heinous
crimes, wartime enemies, hostage taking, and terrorism. Have the lessons of the Crusades and the Inquisition, of the waves of European
imperialism and colonialism, faded from memory—periods when popes
and monarchs, clergy, civil servants, and soldiers often legitimated their
actions in the name of religion: God and country, crown and cross?

Too often, coverage of Islam and the Muslim world concludes that
there is a monolithic Islam out there somewhere, believing, feeling,
thinking, and acting as one. To speak of "the spokesman of a great
spiritual and ethical religion," perpetuates the image of a monolithic
Muslim world seen through the lens of Iran and the Ayatollah Khomeini. The Ayatollah Khomeini was one spokesman, not *the* spokesman, of Islam. There is no pope of Islam. Seeing all Muslims and all
events in the Muslim world through the prism of Khomeini and revolutionary Iran had profound effects upon American perceptions of Islam
and the Middle East. It often obscured the differences and divisions in
the Muslim world—the many countries and Muslims that did not follow
his lead; Iran's failure or at best very limited success in exporting and
inciting Islamic revolutions; and the many voices other than Khomeini's
who spoke out on Salman Rushdie.

The creation of an imagined monolithic Islam leads to a religious reductionism that views political conflicts in the Sudan, Lebanon, Kosovo, Yugoslavia, and Azerbaijan in primarily religious terms—as "Islamic–Christian conflicts." Although the communities in these areas may be broadly identified as Christian and Muslim, it is nonetheless true, as with Northern Ireland's Catholic and Protestant communities, that local disputes and civil wars have more to do with political issues (e.g., ethnic nationalism, autonomy, and independence) and socioeconomic issues than with religion.

Visions of Pan-Islam

Both in the past and today, fear of a monolithic Islamic threat has often been expressed in Pan-Islamic terminology. Triumphant colonial Europe "saw all resistance to its domination as a sinister conspiracy. Such a plot could only be inspired by a cruel, Machiavellian spirit. . . . Whenever there was any show of anti-imperialism, even if it was a purely local reaction, pan-Islam was blamed."[26] Despite the fractured nature of the Muslim world, in the early twenties, during the mandate period and its continued European dominance, the tendency to generalize about a monolithic Pan-Islamic world—a "seething" Islamic ferment—remained firmly embedded in the West.

> In its very hour of apparent triumph, Western domination was challenged as never before. During those hundred years of Western conquest a mighty internal change had been coming over the Moslem world. The swelling tide of Western aggression had at last moved the "immovable" East. At last Islam became conscious of its decrepitude and with that consciousness a vast ferment, obscure yet profound, began to leaven the 250,000,000 followers of the Prophet from Morocco to China and from Turkestan to the Congo. . . . Today Islam is seething with mighty forces fashioning a new Moslem world.[27]

If anyone doubted that Pan-Islam was a global threat to the West, a clash between civilizations, a rebellion fed by hatred, they only needed to listen to the warnings of the noted Orientalist Leone Caetani, who, speaking of the effect of World War I, commented:

> The convulsion has shaken Islamic and Oriental civilization to its foundations. The entire Oriental world, from China to the Mediterranean, is in ferment. Everywhere the hidden fire of anti-European hatred is burning. Riots in Morocco, risings in Algiers, discontent in Tripoli, so-called Na-

tionalist attempts in Egypt, Arabia, and Libya are different manifestations of the same deep sentiment and have as their objective the rebellion of the Oriental world against European civilization.

The principal reason for this ferment is the report spread throughout the world that the Entente wishes to suppress the Ottoman Empire, dividing its territory among the powers and ceding Palestine to the Jews.[28]

Fear of a Pan-Islamic, global uprising still persists today. In the wake of the Iranian Revolution, the Ayatollah Khomeini's call for other Islamic revolutions found ready believers not only in the Muslim world but also in the West. In France, Raymond Aron "warned of the Islamic 'revolutionary wave,' generated by 'the fanaticism of the prophet and the violence of the people,' which the Ayatullah Khumayni has unleashed."[29] In 1980 U.S. Secretary of State Cyrus Vance stated that a major reason for his objection to a military mission to rescue American hostages in Iran was fear of an "Islamic–Western war. . . . Khomeini and his followers, with a Shiite affinity for martyrdom, actually might welcome American military action as a way of uniting the Moslem world against the West."[30] A decade later, with the collapse of communism, Charles Krauthammer wrote: "History is being driven by another force as well: the political reawakening of the Islamic world."[31] It is a challenge all the more ominous because it is Pan-Islamic. It is a "global intifada," embracing not only the Islamic heartland but also the peripheries of the Muslim world where Islam confronts the non-Muslim communities—in Kashmir, Azerbaijan, Kosovo in Yugoslavia, Lebanon, and the West Bank.

Krauthammer's piece reflects the monolithic/reductionist/threat approach to the Islamic world. Fundamentalism is equated with "fundamentalist Koran-waving Khomeniism."[32] Despite his recognition of multiple motives (religious and nonreligious) and differing political contexts, Krauthammer imposes a unity by asserting that "a deep historical current runs through" these random explosions, by speaking of a "pan-Islamic demand," and by emphasizing the shared political thread and geographical unity of this "global intifada." The use of these terms yields an imagined rather than an empirical transnational Islamic unity and reality.

Talk of a "global" intifada also distracts and detracts from the real nature of discontent in the Palestinian intifada, by implying that it is simply part of a transnational Islamic uprising rather than an Arab–Israeli problem. The intifada is first and foremost an uprising of Arab Muslims and Christians. The primary cause of the intifada was not Islam or Islamic revivalism but continued Israeli occupation of the West Bank and Gaza and the desperation of young Palestinians in particular. The Islamic dimension (Hamas and Islamic Jihad) of the confrontation increased significantly as the Palestinian leadership proved unable to effectively resolve the situation. Moreover, the phrase "global intifada" imposes a superficial unity on events that occur in quite diverse contexts and obscures the profound differences between Kashmir and Kosovo, Azerbaijan and Lebanon.

It seems ironic that while applauding the change of governments and political turmoil in Eastern Europe, the Balkans, and the Caucasus, which resulted from the collapse of communism, Krauthammer views Muslim demands for self-determination as a cause for concern. Krauthammer is in the curious position of raising the alarm over what he identifies as essentially twentieth-century independence movements in the Muslim world. The first phase occurred during the first half of the twentieth century, when the Islamic heartland "[f]rom Morocco to Pakistan . . . threw off European imperialism in a process that began earlier in the century and may be said to have culminated with the revolution in Iran." The second phase is "the further evolution of the Islamic awakening: the demand for local hegemony by Moslem populations at the borders of the Islamic world." Are freedom and self-determination or the use of violence to regain one's independence and sovereignty less valid and a threat if undertaken in the name of religious or ethnic nationalism? Is a Western secular profile (liberal secular nationalism) a requirement for freedom and acceptance within the international community?

Monolithic Islam has been a recurrent Western myth which has never been borne out by the reality of Muslim history. When convenient, Western commentators waste little time on the divisions and fratricidal relations of the Arab and Muslim world so as to underscore its intractable instability. Sen. Albert Gore, speaking of Syrian–Iraqi relations, noted: "Baathite Syrians are Alawites, a Shiite heresy, while the Iraqis are Sunnis. Reason enough in this part of the world for hatred and murder."[33] Yet when equally convenient, Islam, the Arabs, and the Muslim world are represented as a unified bloc poised against the West. However much the ideals of Arab nationalism or Islam speak of the unity and identity of a transnational community, history has proven otherwise. At best, some Muslims, as in the Iranian Revolution, have achieved a transient unity in face of a common threat, a solidarity which dissipates as easily as it was formed, once the danger has subsided and competing interests again prevail. The inability of Arab nationalism, Arab socialism, Iran's Islamic Republic, or Muslim opposition to the Soviet Union's invasion of Afghanistan to produce a transnational or regional unity, as well as the disintegration of the Arab coalition (Iraq and the Gulf states) against Iran after the Iran–Iraq war, are but several modern examples. As James Piscatori has observed, "The problem with assuming a unified response is that it conceals the reality of . . . entrenched national differences and national interests among Muslims."[34]

Diversity rather than Pan-Islamic political unity is also reflected in foreign policy. The common "Islamic" orientation or claim of some governments reveals little unity of purpose in interstate and international relations because of conflicting national interests or priorities. Qaddafi was a bitter enemy of Sadat and Nimeiri at the very time when all three were projecting an "Islamic" image. Khomeini's Islamic Iran consistently called for the overthrow of the house of Saud on Islamic grounds, their rivalry even erupting during the annual pilgrimage to Mecca. Islamically identified governments also reflect differing relationships with the West. While Libya and Iran's relationship with the West, and with the United States in particular, has often been confrontational at the same time, the United States has had strong allies in Saudi Arabia, Egypt, Kuwait, Pakistan, and Bahrain. National interest and regional politics rather than ideology or religion remain the major determinants in the formulation of foreign policy.

Democratization

Integral to our view of the Islamic threat is the belief that Islam is inherently antidemocratic and intolerant. In the wake of the Gulf War of 1991, the issue of political liberalization and democracy reemerged in the Middle East and in Western diplomacy.[35] Democracy has been slow to come to the Middle East. Despite Western influence and a façade of parliamentary government in some countries, the political reality has more often been one of authoritarianism. Political parties are banned or severely restricted; elections are often rigged. Of the states in the Middle East, six are monarchies and seven are dictatorships; Algeria, Egypt, and Tunisia have historically been dominated by one party.

Democracy movements and pressures upon Muslim governments for greater liberalization predated the Gulf War. As the Soviet Union and Eastern Europe were swept along by the wave of democratization in 1989–90, the demands of Muslim nationalities in the Soviet Union for greater autonomy, the Palestinian intifada, and Kashmiri Muslim demands for independence from India captured the attention of many in the Muslim world. Secular and Islamic activists increasingly couched criticisms of their regimes in the language of political liberalization and democracy. As we have seen in chapters 4 and 5, in many Muslim countries Islamic activists not only took to the streets but also turned to the ballot box. Both secular and Islamic organizations and political parties pressed for political reforms with limited success: elections in Egypt, Tunisia, Algeria, Jordan, and Pakistan, and, in the aftermath of the Gulf War, a promised election in Kuwait and the resurrection of the pledge of a consultative assembly in Saudi Arabia.

Talk of democratization troubles both autocratic rulers in the Muslim world and many Western governments. The former fear any opposition, all the more so one that cloaks itself in values that Western governments officially cherish and preach. For leaders in the West, democracy raises the prospect of old and reliable friends or client states being transformed into more independent and less predictable nations which might make Western access to their oil less secure. Thus stability in the Middle East has often been defined in terms of preservation of the status quo.

Lack of enthusiasm or support for political liberalization in the Middle East has been rationalized by the claim that both Arab culture and Islam are antidemocratic. The proof offered is the lack of a democratic tradition—more specifically, the paucity of democracies in the Muslim world. Why the glaring absence of democratic governments? The political realities of the Muslim world have not been conducive to the development of democratic traditions and institutions. European colonial rule and postindependence national governments headed by military officers, monarchs, and ex-military rulers have contributed to a legacy which has had little concern for political participation and the building of strong democratic institutions. National unity and stability as well as political legitimacy have been undermined by the artificial nature of modern states whose national boundaries were often determined by colonial powers and whose rulers were either placed on their thrones by Europe or seized power for themselves. Weak economies, illiteracy, and high unemployment, especially among the younger generation, exacerbate the situation, undermining confidence in governments and increasing the appeal of "Islamic fundamentalism."

Islam and Democracy

Are Islam and democracy necessarily or inherently incompatible? The democratization movement in the Middle East and the participation (and successes) of Islamic movements in electoral politics raise the question of the compatibility of Islam and democracy.[36] Among the arguments proffered by those who fear the promotion of a democratic process in the Muslim world is that it risks the "hijacking of democracy" by Islamic activists and further Islamic inroads into centers of power, threatening Western interests and fostering anti-Westernism and increased instability.

> Today in most Islamic countries, free elections would produce fundamentalist victories and validate the imposition of theocracy. . . . [Fundamentalists] have pressed for free elections in several Arab countries. Presenting themselves as the protectors of the oppressed, they have done quite well in these elections as they knew they would. But it is questionable that their real aim is to promote democracy. . . . Islam draws no line between religion and politics. As undemocratic as the present Saudi regime is, a total Islamic one—even with broader political participation—would be less free. It would have no neutral public space where people's views are treated as opinions, not as truth. Elections would become trivial in that environment.[37]

Many argue that Islamic values and democratic values are inherently antithetical, as seen in the inequality of believers and unbelievers as well as of men and women.

History has shown that nations and religious traditions are capable of having multiple and major ideological interpretations or reorientations. The transformation of European principalities, whose rule was often justified in terms of divine right, into modern Western democratic states was accompanied by a process of reinterpretation or reform. The Judaeo-Christian tradition, while once supportive of political absolutism, was reinterpreted to accommodate the democratic ideal. Islam also lends itself to multiple interpretations; it has been used to support democracy and dictatorship, republicanism and monarchy. The twentieth century has witnessed both tendencies.

Some leaders of Islamic movements have spoken out against Western-style democracy and a parliamentary system of government. Their negative reaction has often been part of the general rejection of European colonial influence, a defense of Islam against further dependence on the West rather than a wholesale rejection of democracy. In recent decades many Muslims have accepted the notion of democracy but differed as to its precise meaning. The Islamization of democracy has been based upon a modern reinterpretation of traditional Islamic concepts of political deliberation or consultation (*shura*), community consensus (*ijma*), and personal interpretation (*ijtihad*) or reinterpretation to support notions of parliamentary democracy, representative elections, and religious reform. While radical revolutionaries reject any form of parliamentary democracy as Westernizing and un-Islamic, many Islamic activists have "Islamized" parliamentary democracy, asserting an Islamic rationale for it, and appealed to democracy in their opposition to incumbent regimes.

Islamic organizations such as the Muslim Brotherhoods in Egypt, the Sudan, and Jordan, the Jamaat-i-Islami in Pakistan, Kashmir, India, and Bangladesh, and Algeria's Islamic Salvation Front, Tunisia's Renaissance Party, Kuwait's Jamiyyat al-Islah (Reform Society), and Malaysia's ABIM and PAS, among others, have advocated the principle of democratic elections and, where permitted, have participated in parliamentary elections.

There are differences between Western notions of democracy and Islamic traditions. Increased emphasis on political liberalization, electoral politics, and democratization does not necessarily imply uncritical acceptance of Western forms of democracy. A commonly heard argument is that Islam possesses or can generate its own distinctive forms of democracy in which popular sovereignty is restricted or directed by God's law. Thus both divine and popular sovereignty are affirmed in a delicate balance capable of producing multiple forms and configurations.

Electoral successes of Islamic candidates in Algeria, Egypt, Jordan, Tunisia, and the Sudan feed the fears of nervous rulers. In a climate in which most governments since 1979 have been traumatized by the specter of "Muslim fundamentalism," this fear has provided an excuse for continuing to limit political liberalization and democratization. At best the attitude of many rulers may be characterized, in the words of one Western diplomat, as an openness to "risk-free democracy"! Both the Tunisian and Algerian governments' management of political liberalization reflect this approach. Openness to government-controlled change—yes; openness to a change of government that would bring Islamic activists to power through democratic means—no. Opposition parties and groups are tolerated as long as they remain relatively weak or under government control and do not threaten the ruling party.

Democracy has become an integral part of modern Islamic political thought and practice. It has become accepted in many Muslim countries as a litmus test by which both the openness of governments and the relevance of Islamic groups are certified. It is a powerful symbol of legitimacy, legitimizing and delegitimizing precisely because it is seen to be a universal good. However, questions as to the specific nature and degree of popular participation remain unanswered. In the new Muslim world order, Muslim political traditions and institutions, like social conditions and class structures, continue to evolve and are critical to the future of democracy in the Middle East.

A major issue facing Islamic movements is their ability, if in power, to tolerate diversity. The status of minorities in Muslim-majority areas and freedom of speech remain serious issues. The record of Islamic experiments in Pakistan, Iran, and the Sudan raises serious questions about the rights of women and minorities under Islamically oriented governments. The extent to which the growth of Islamic revivalism has been accompanied in some countries by attempts to restrict women's rights, to enforce the separation of women and men in public, to require veiling, and to restrict their public roles in society strikes fear in some segments of Muslim society and challenges the credibility of those who call for Islamization of state and society. The record of discrimination against the Bahai in Iran and the Ahmadi in Pakistan as "deviant" groups (heretical offshoots of Islam), against Christians in the Sudan, and Arab Jews in some countries, as well as increased sectarian conflict between Muslims and Christians in Egypt and Nigeria, pose similar questions of religious pluralism and tolerance.

If many Muslims ignore these issues or facilely talk of tolerance and human rights in Islam, discussion of these questions in the West is often reduced to two contrasting blocs; the West which preaches and practices freedom and tolerance, and the Muslim world which does not. Muslim attitudes toward Christian minorities and the case of Salman Rushdie are marshaled to support the indictment that Islam is intolerant and antidemocratic.

Muslim demands for independence are regarded as deceptive and a threat to minorities.

> What is being pursued, therefore, is not Wilsonian self-determination (though many of these intifadas have adopted its language), because in the Islamic world self-determination is permitted only to Moslems. What instead is being pursued is a pan-Islamic demand for sovereignty over any territory where Moslems form a local majority.[38]

The charge of intolerance is supported by statements from some Christian church leaders, quoted in syndicated columns such as that of Patrick Buchanan, who claim that Muslim fundamentalists have no use for ecumenism or dialogue, which is a "betrayal to Allah. . . . For Islam, there is only one revelation, the final word has been spoken in the Quran, and Muhammad is the final Prophet."[39] This assertion is misleading and inaccurate. As we have seen earlier, Islam acknowledges God's revelation to Jews and Christians, just as it does biblical prophets, though it considers the Quran to be the final and complete revelation and Muhammad the last in a long line of prophets. Following the logic of the preceding statement, Christianity could present the same problem. Christians have traditionally believed that they received the final, complete revelation, possess not just a prophet but the Son of God, and have a universal mission to convert the world. Anyone who has participated in local, national, and international ecumenical gatherings knows that Muslims are regular participants. This does not negate the fact that some Muslims and Islamic organizations, like some Christians and Jews, have little use for ecumenism. While there are Muslims, Jews, and Christians whose religious positions produce an inflexible blend of religion and self-righteousness, many, while affirming the truth of their own faith, are increasingly open to dialogue with other believers as they experience the realities of a pluralistic and interdependent world.

Yet the status of non-Muslims and the implications of political pluralism remain contemporary Islamic issues. Without a reinterpretation of the classical Islamic legal doctrine regarding non-Muslim minorities as "protected people" (*dhimmi*), an ideologically oriented Islamic state would be at best a limited democracy with a weak pluralistic profile; its ideological orientation would restrict the participation of non-Muslims in key government positions, and the existence of political parties representing a competing ideology or orientation: secular, communist, socialist. Non-Muslims would be second-class citizens with limited rights and opportunities, as is now the case in countries such as Iran and Pakistan.

Despite democratic tendencies in the Muslim world and among Islamic activists, multiple and conflicting attitudes toward democracy continue to exist and so leave the future in question. Thus for example while Abbasi Madani, the leader of Algeria's Islamic Salvation Front, affirmed his acceptance of democracy in the face of accusations that he had opposed the democratic process in the past, some of the Front's younger voices like the popular preacher Ali Benhadj rejected democracy as an un-Islamic concept.[40] Only time will tell whether the espousal of democracy by many contemporary Islamic movements and their participation in the electoral process are simply a means to power or a truly embraced and internalized goal resulting from religious reinterpretation informed by both faith and experience. Based on the record thus far, one can expect that where Islamic movements come into power—as is already the case with many governments in the Middle East, secular as well as Islamic—issues of political pluralism and human rights will remain sources of tension and debate. Greater political liberalization and participation are part of a process of change that requires time and experience to develop new political traditions and institutions. However, if attempts to participate in the electoral process are blocked, crushed, or negated as in North Africa, the currency of democracy as a viable mechanism for political and social change will be greatly devalued in the eyes of many.

The Rushdie Affair

Just as tolerance and freedom are equated with the West, so too is liberalism, while illiberalism is imputed to Islam. The Salman Rushdie case is often cited to illustrate that "the followers of Muhammad, even in the West, have little use for the liberalism of J. S. Mill."[41] Such judgments reduce the complexity of the Rushdie uproar to the issue of Islam's incompatibility with Western liberalism, rather than the question of whether religious belief places limits on free speech; it also implies that all Muslims hold a single position or speak with one voice.

Any discussion of the Rushdie affair must be seen against the background of the past, must recognize the historical and international context in which the debate occurred. As John Voll noted, the complex history of Islam and the West,

> [a] long history of frequently hostile relations between Muslim and Western societies . . . involves the development on both sides of traditions of conscious defamation of "the enemy" as a part of the rivalry. Thus, we have to overcome a heritage of misunderstanding and prejudice that is deeply rooted in society. The "holy war mentalities" of Crusade and jihad in this heritage color many positions in the debate over *The Satanic Verses*. American commentators speak of the threat to constitutional values in the United States, and Westerners often portray the Ayatollah as engaging in a holy war against Western values of tolerance and freedom of expression. At the same time, many Muslims view Mr. Rushdie's book as being in the Western tradition of defamation of Islam and as part of the West's Crusade against basic Islamic values of community responsibility and obedience to God.[42]

It is of equal importance that, while many Muslims repudiated Rushdie for blasphemy, all did not declare him an apostate or summarily condemn him to death. Muslim reactions varied, though many Western commentators and the media often ignored that fact. The multiple and diverse voices in the Muslim world were drowned out by more strident Western and Muslim voices and the fixation on Khomeini which dominated media coverage.

The furor over Salman Rushdie's *Satanic Verses* quickly became an international event.[43] The book was condemned and burned by Muslims in many parts of the world. Orderly demonstrations and protest were overshadowed by crowds in the streets of Teheran calling for the death of the author, a demonstration in front of the American Center in Islamabad, Pakistan, on February 12, 1989, which cost the lives of at least five Pakistanis, and riots and threats against the book's publishers and bookstores in Britain and the United States. The turmoil surrounding the book led to a further cooling of relations between Iran and the West, as European nations condemned the Ayatollah Khomeini's call for the execution of Rushdie and withdrew diplomats from Iran.

Muslims were offended by passages in the book which questioned the authenticity of the Quran (*The Satanic Verses,* p. 367), ridiculed the Prophet and the contents of the Quran (pp. 363–64), and referred to Muhammad as "Mahound"—a term used in the past by Christian authors to vilify Muhammad (p. 381). The book also had prostitutes assuming the identity and names of Muhammad's wives, and the very Quranic symbol for their seclusion and protection, "The Curtain," is transformed into the image of a brothel, which men circumambulate as worshipers do the sacred shrine (Kaaba) during the pilgrimage to Mecca (p. 381).

Muslim responses varied.[44] Most denounced the book as an attack upon Islam. Many Muslims in the Muslim world and the West accused the author of blasphemy and sought to have *The Satanic Verses* banned. The Secretary-General of the Organization of the Islamic Conference (OIC) urged all forty-five member nations to ban the book and boycott the publishers. Many Muslim countries, as well as India, which has a significant Muslim minority, banned the book. Egypt had quietly done so in November 1988.

On February 14, 1989, the Ayatollah Khomeini issued a *fatwa* (religious decree or legal opinion) which condemned Rushdie, who was born a Muslim but left the faith, to death and called for his execution. The Mufti of Sokoto (Nigeria) also called for Rushdie's death. In contrast, religious scholars at Cairo's al-Azhar University, the oldest university in the world and a venerable center of Islamic learning, stated that, according to Islamic law, Rushdie must first have a trial and be given an opportunity to repent. The Egyptian Nobel laureate Naguib Mahfouz, a strong proponent of freedom of expression, publicly backed Egypt's ban, stating that "different cultures have different attitudes towards freedom of speech. What might be endured in Western cultures might not be acceptable in Muslim countries."[45]

The Rushdie affair also had important policy implications. In the aftermath of the Iran–Iraq cease-fire in August 1988, Iran had taken notable steps to improve its image and relations internationally. Diplomatic and trade relations with Europe had been strengthened, and Khomeini had sent a personal message to the Soviet leader, Mikhail S. Gorbachev, exhorting him to look to Islam instead of Marxism for a solution to his country's problems. The Iranian government had been taking steps to improve its image in general and the perception of its human rights policy in particular, and had hosted and participated in a number of conferences involving Europeans and Americans.

However, the more hard-line, anti-Western faction within Iran was able to use the Rushdie affair to interrupt this turn toward normalization of relations with the West. It had found the right button to press in motivating Khomeini to respond as Iran's supreme religiopolitical leader. Thus he condemned the book and the author, denounced "the devilish acts of the West," and approved Iran's severance of diplomatic relations with Britain.

Memories of the American embassy hostages in Iran, of the burning down of the American embassy in Pakistan in 1979, of attacks against Americans in Lebanon and other countries in the aftermath of the Iranian Revolution, and of the riot outside the American Center in Islamabad over the Rushdie affair, are vivid reminders of the treacherous pitfalls in relations between the United States and militant activists. In contrast to some European governments, the American administration responded more slowly and carefully. It distanced itself from the book (Vice President Quayle described it as offensive), while affirming the right of free speech and rejecting Khomeini's call for the assassination of the author.

One of the lessons of the Rushdie affair is the awareness that differences do exist in worldview and values between Western and Muslim cultures. The West's modern secular orientation can blind it to the reality to which many Muslims reacted, an act which they regard as an attack on the most sacred bases of their faith. It would be equally unfortunate to lose sight of the complexity of Islam, of the multiple voices and diverse motives in the Muslim community and politics. The Ayatollah Khomeini was not the sole spokesperson of Islam, nor was it useful to place all Islamic issues within the memory and emotions of the past decade's contentious relationship with Iran. For many Muslims the Rushdie affair was primarily a religious matter. Yet for others in countries like Iran and Pakistan, religious outrage was often conveniently wedded to domestic politics, such as elite factionalism in Iran or anti–Benazir Bhutto sentiments in Pakistan. In the United Kingdom the Rushdie affair was part of the larger questioning of a society that many Muslims believed did not seem to live up to its own standards for tolerance religiously and politically. For example, Muslims have repeatedly pointed out that the law against blasphemy protects only the Christian faith—in effect, the social and political establishment—and does not protect other religions like Judaism and Islam. As Mashuq Ally of the University of Wales noted, "The Muslim community in this case saw itself as fighting a battle of principle: the injustice of the fact that the law of blasphemy applied only to a particular group within the society and not to other groups."[46] Both an insensitivity to cultural differences, and failure as well to appreciate differing political contexts, limited the ability to understand the depth of Muslim anguish and yes, rage. Acknowledgment of the offensiveness of *The Satanic Verses* to many Muslims does not require agreement, but it can include recognition that the limits of free speech may vary in different religio-cultural contexts.

Islam and the Gulf War of 1991

If Saddam Hussein had had his way, Iraq would have emerged from the Gulf War as the defender of Islam against Western imperialism. However, the war again undermined stereotypical images of monolithic Arab and Muslim worlds gripped by Pan-Arabism or Pan-Islam. The Gulf crisis of 1990–91 divided the Arab and indeed the Muslim worlds.[47] Similarly, it witnessed multiple appeals to Islam by Muslim political and religious leaders to legitimate each side in the conflict and tested the ideology and allegiances of Islamic movements.[48] Perhaps nothing seemed more incongruous than that Saddam Hussein, the head of a secularist regime who had ruthlessly suppressed Islamic movements at home and abroad, would cloak himself in the mantle of Islam and call for a jihad.

The Gulf crisis simultaneously presented an apparent united Arab response to a rapacious, expansionist Iraq and, at a deeper level, an Arab and indeed Muslim world divided to an unparalleled extent. The Arab League and the Organization of the Islamic Conference, a Saudi-supported organization of forty-four Muslim states, condemned Saddam's invasion of Kuwait. Emphasis on Arab and Muslim government support for the alliance against Saddam Hussein obscured deeper divisions. Only twelve of the Arab League's twenty-one members supported the anti-Saddam forces. Moreover, as the crisis dragged on, Saddam came to enjoy a degree of popular sympathy often not fully appreciated in the West, where the tendency was to equate the position of Arab and Muslim governments with their people. Little distinction was made between the differing perspectives of the Western-led coalition supported by their Arab and Muslim allies, and the views of a significant portion of the populace whose deep-seated grievances were given a new voice and champion in Saddam Hussein.

Stunned by the unexpectedly quick and broadly based international condemnation, Saddam increasingly emphasized the Arab and Islamic rationales for his actions and thus created popular pressure on Arab and Muslim rulers. He stepped into a leadership vacuum in the Arab world.

Although not a charismatic leader, he created a popular persona by appealing to Arab nationalism and Islam, shrewdly exploiting long-standing issues: the failures of Arab governments and societies (poverty, corruption, maldistribution of wealth), the plight of the Palestinians, and foreign intervention leading to Arab dependency. Like the Ayatollah Khomeini, Saddam appealed to Islam to enhance his image as the champion of the Palestinians, of the poor and oppressed, and as the liberator of the holy places, as well as to legitimate his call for a holy war against the Western (especially U.S.) occupation of Arab lands and control of Arab oil, and to overthrow those Arab regimes that opposed him. The espousal of populist causes—issues that transcend the individuals who from time to time champion them—transformed Saddam the secular despot into a popular hero among many; the messenger was transformed by the message.

The deep divisions within the Muslim world were reflected in competing appeals to Islam and calls for jihad. Saudi Arabia and Egypt's leading religious leaders legitimated the presence of foreign troops in Saudi Arabia, the home of Islam's holy sites (Mecca and Medina), access to which is forbidden to non-Muslims. At the same time Saddam Hussein, Iran's Ayatollah Khamenei, and the Jordanian *ulama* and Muslim Brotherhood called for a jihad against foreign intervention.[49]

Islamic movements, reflecting their societies, were initially pulled in several directions. At first, most condemned Saddam Hussein, the secular persecutor of Islamic movements, and denounced his invasion of Kuwait. However, even though many movements had supporters in Saudi Arabia, when large numbers of foreign troops poured into the Gulf, their initial rejection of Saddam gave way to a more populist, Arab nationalist, anti-imperialist support for him, and to a condemnation of foreign intervention and "occupation" of Islam's homeland, Saudi Arabia. The key catalyst was the massive Western (especially U.S.) military buildup in the region and the threat of military action and a permanent Western presence.

For many in the Muslim world, the buildup of foreign troops, announced after the American elections in November 1990, transformed the nature of the conflict from defensive to offensive. Operation Desert Storm had been transformed into Operation Desert Sword; the defense of Saudi Arabia and the liberation of Kuwait had become not just a war against Saddam Hussein but an all-out attempt to destroy Iraq politically and militarily. Who would benefit most from the resulting power vacuum? A common answer was: America and its ally Israel.

Domestic politics—pressure not to run counter to popular sentiment—

as much as religious conviction and ideology influenced Islamic activists in their support for Saddam and his call for jihad, as witnessed in Algeria, Tunisia, Morocco, Jordan, Pakistan, and Egypt. Initially thousands of Muslim activists in Algeria had demonstrated against Iraq's invasion of Kuwait; but on a subsequent visit to Baghdad, Abbasi Madani, the leader of the Islamic Salvation Front, declared: "any aggression against Iraq will be confronted by Muslims everywhere."[50] In Jordan the Muslim Brotherhood had also initially condemned the Iraqi invasion. After the deployment of American forces, however, it called for a jihad against "the new crusaders in defense of Iraq and the Islamic world." As one American Muslim observer noted: "people forgot about Saddam's record and concentrated on America. . . . Saddam Hussein might be wrong, but it is not America who should correct him."[51] In Egypt the Muslim Brotherhood first denounced the Iraqi invasion, then supported the government's anti-Saddam position. As the war continued, the Brotherhood joined with other opposition groups and criticized the massive presence of Western forces in Saudi Arabia.

Even in countries that sent forces to support the anti-Saddam "international alliance," popular sentiment often differed from that of the government. In a poll taken by a Pakistani magazine, the *Herald,* where a majority had opposed the annexation of Kuwait, 86.6 percent of those polled responded negatively to the question, "Should U.S. troops be defending the Muslim holy places in Saudi Arabia?"[52]

In Southeast Asia, Malaysian and Indonesian government support for U.N. resolutions was accompanied by general condemnation of Iraqi aggression in Kuwait. However, this did not translate into popular support for foreign troops or the threat of massive military action in the Gulf. Saddam Hussein's aggression and the Bush administration's portrayal of him as a new Hitler were not isolated, in the minds of many, from the turbulent politics of the Middle East or the obvious self-interest of the Western-led coalition and many of its allies. Defending Saudi Arabia and liberating Kuwait was one thing, but attacking Iraq quite another. Many in Malaysia and Indonesia (reflecting, or in solidarity with, populist sentiment throughout much of the Muslim world), while critical of Saddam's annexation of Kuwait, were equally strong in their condemnation of America's "double standard": excoriating Saddam for violating international law and vigorously enforcing the U.N. resolutions, while refusing to take the very same stand with regard to U.N. resolutions condemning Israel's continued occupation of the West Bank and Gaza.[53] An independent Malaysian publication reflected the strong feelings of many in the Muslim world:

Given the magnitude of the U.S. military buildup in the Gulf and its past records of invasions of and interferences in Third World countries, it is likely that the situation in the Gulf will lead to full-scale war. . . . It is indeed hypocritical for the U.S. to come to the aid of Kuwait while it remains silent about Israel's invasion and occupation of the West Bank, Gaza Strip, Golan Heights, Lebanon and its bombing of Tyre, Sidon, and West Beirut, which wounded and killed hundreds of civilians. Instead of economic sanctions, Israel received an increasing amount of U.S. aid. At present, Israel is receiving $4 billion a year from the U.S. $400 million of U.S. aid goes to help settle Jews from the Soviet Union in Israel's occupied territories, therefore legitimising Israel's territorial expansion. In addition, the U.S. had consistently vetoed all attempts by the Security Council to condemn Israel for ignoring United Nations' demands that it withdraw from Lebanon. This double standard was again observed by the U.S. in regard to the Palestinian intifada in which the Israeli military killed over 700 Palestinians including some 160 children. The UN resolution calling for international observers to investigate the situation in the occupied territories was vetoed by the U.S.[54]

The United States's reluctance to link (or to acknowledge the linkage of) the two issues in resolving the Gulf crisis was seen by many in the Muslim world as an attempt to disengage two already interlocked realities.[55]

In Pakistan Qazi Hussein Ahmad, the leader of the Jamaat-i-Islami and a member of Pakistan's senate, while addressing a large anti-American rally called upon the government to give up its pro-American foreign policy. Characterizing the American–European alliance as "anti-Islamic forces," he warned that the U.S. intention was not to liberate Kuwait but to establish its hegemony in the region. Europe and America had joined hands to "destroy the fighting power of the Islamic world."[56] Thus, he concluded, there would be no place for the Islamic world in the New World Order.

Populist support for Saddam Hussein became more strident with pro-Saddam and anti-American rhetoric, demonstrations, protest marches, and at times violence across the Muslim world: in Algeria, Tunisia, Morocco, Mauretania, the Sudan, Libya, Nigeria, Lebanon, Jordan, the West Bank and Gaza, Yemen, Egypt, India, Pakistan, Bangladesh, Iran, Malaysia, Indonesia, and even among South Africa's Muslim community. Scud missile attacks on Tel Aviv were celebrated by many Arab nationalists and Islamic activists alike. Saddam was seen as the first Arab leader to strike directly at the heart of Israel. Those convinced that Israel supplied intelligence information on Iraq, and aware that it had fought its wars with the Arabs on Arab territory and freely carried out retaliatory attacks against Tunis, Baghdad, Beirut, and southern Lebanon, hailed Saddam for bringing war home to the Israelis.

Algeria, Morocco, Tunisia, Mauretania, and the Sudan witnessed large demonstrations. Saddam's appeals to Arab nationalism, Islam, and the Palestinian cause resonated with memories of French colonialism and the struggle of many to root their identity more indigenously in an Arab–Islamic past. In Rabat on February 3, 1991, three hundred thousand took to the streets demanding the withdrawal of Moroccan troops from the multinational coalition. Lebanon experienced some of the largest demonstrations since the beginning of the Lebanese civil war, along with bombings of Western banks and airline offices. In Iran anti-American demonstrations took place in many major cities. President Rafsanjani, while carefully maintaining Iran's official position of neutrality, with equal care aligned himself with populist Muslim sentiment both within Iran and throughout the Muslim world: "the leader of imperialism is destroying a Muslim nation and ruining the resources of Muslims."[57] The Indonesian government supported U.N. resolutions on the Gulf, and many Indonesians denounced the annexation of Kuwait. However, Saddam's linkage of his actions with resolution of the Palestinian problem, the presence of U.S. forces in the Gulf, and the subsequent devastation of Iraq produced a growing level of anti-American criticism in the press and among intellectuals, Muslim leaders, and student groups.[58]

Coalition members such as Morocco, Bangladesh, Egypt, Syria, and Pakistan, who had sent troops to the Gulf, were increasingly subjected to domestic pressure to remove their forces. Islamic activists joined with other opposition groups in protesting the war. In Egypt the Muslim Brotherhood joined with the Left and many professional associations. An antigovernment weekly carried the headline: "Muslims! Your brothers are being annihilated. Hurry to aid Iraq in its heroic steadfastness!"; it sold out within hours.[59] Pakistan saw religious organizations like the Jamaat-i-Islami involved in massive public demonstrations throughout the country as hundreds volunteered for the jihad.

As in other instances, the Gulf War revealed a diversity of positions among Muslim governments, between some governments and their people, and among religious leaders and Islamic movements.

The Islamic Threat: Issues of Interpretation

The ways in which we understand the nature of religion and the relationship of religion to politics and society greatly determine our expectations and judgments. Why has the continued vitality of Islamic revivalism been underestimated, and why does it continue to be primarily per-

ceived and responded to as a threat? More than ten years have elapsed
since the Iranian revolution and the West's interest in, if not fear of,
Islamic revivalism or fundamentalism. If Islam was insufficiently reported on, and if ignorance or lack of awareness characterized the past,
many would argue that in recent years Islamic revivalism and Islamic
politics have become a growth industry.

Despite greater interest and media coverage and a seemingly never-
ending stream of publications, understanding and appreciation of Islam
and Islamic movements have remained limited and selective. Given the
resources available, the number of academic experts and government
analysts, why did it take the Iranian Revolution of 1978–79 to draw
attention to events that had been taking place throughout the seventies
in Libya, Egypt, Pakistan, the Sudan, and the broader Muslim world? In
spite of the post factum "discovery" of Islamic revivalism, why did most
analysts continue to underestimate its vitality and thus fail to appreciate
its power and growth in countries like Algeria, Tunisia, and Jordan?
Conversely, why has there been a persistent tendency to reduce Islam
and Islamic activism to religious extremism and terrorism? Violence,
terrorism, and injustice exist in the Muslim world as in other areas of the
world. They have been legitimated on occasion in the name of Islam, as
likewise in the name of Christianity, Judaism, and secular ideologies
such as democracy and communism.

At the heart of Western misinterpretation, stereotyping, and exaggerated fears of Islam is a clash of viewpoints. The stereotypes of the
"Arab" and Islam in terms of Bedouin, desert, camel, polygamy, harem,
and rich oil shaykhs have been replaced by those of gun-toting mullahs
or bearded, anti-Western fundamentalists. Western fears and antipathy
are fed not only by media reports and recent events but also by an
outlook on life which is often antithetical to that of Islamic activists.
Modern, post-Enlightenment secular language and categories of thought
distort understanding and judgment. The modern notion of religion as a
system of personal belief makes an Islam that is comprehensive in scope,
with religion integral to politics and society, "abnormal" insofar as it
departs from an accepted "modern" norm, and nonsensical. Thus Islam
becomes incomprehensible, irrational, extremist, threatening.

What most forget is that all the world's religions in their origins and
histories were fairly comprehensive ways of living. While the relationship
of religion to politics has varied, religion is a way of life with a strong
emphasis on community as well as personal life: the way of the Torah, the
straight path of Islam, the middle path of the Buddha, the righteous way
(dharma) of Hinduism. They provide guidance for hygiene, diet, the
managing of wealth, stages of life (birth, marriage, death), and ritual and

worship. The modern notion of religion has its origins in the post-Enlightenment West. Its restricted definition has become accepted as the norm or meaning of religion by many believers and unbelievers alike in the West. Bereft of a sense of history, few realize that the term *religion* as known and understood today is modern and Western in origins. Similarly, it was the West that then set about naming other religious systems or isms. Christianity and Judaism were joined by the newly named Hinduism, Buddhism, and Mohammedanism. Thus the nature and function of other religious traditions were categorized, studied, and judged in terms of modern, post-Enlightenment, secular criteria, with its separation of Church and State.

As most forget or are ignorant of the fact that our concept of religion is a modern construct, so too there is a tendency to forget that the Western notion of separation of Church and State is relatively new. Historically, the dividing line between faith and politics among the Abrahamic faiths (Judaism, Christianity, and Islam) had been blurred from the biblical conquests and the early Jewish kingdoms to imperial Christianity. As the Christian empires and Crusades demonstrated, while Church and State were distinct, in Christianity they were not always separate. However, modern notions of religion as a system of belief for personal life, and separation of Church and State, have become so accepted and internalized that they have obscured the beliefs and practice of the past and come to represent for many a self-evident and timeless truth. As a result, from a modern secular perspective the mixing of religion and politics is regarded as abnormal, dangerous, and extremist. Thus when secular-minded peoples (government officials, political analysts, the mass of the general public) in the West encounter Muslim individuals and groups who speak of Islam as a comprehensive way of life, they immediately dub them "fundamentalist" with the connotation that these are backward-looking individuals, obstacles to change, zealots who are a threat. The attitude of many governments and secular elites in the Muslim world is often similar. Images of militant mullahs and the violent actions of some individuals and groups are then taken as representative and proof of the inherent danger of mixing religion and politics.

Viewing Islam and events in the Muslim world primarily through the prism of violence and terrorism has resulted in a failure to see the breadth and depth of contemporary Islam. In particular, many have overlooked the quiet revolution that has occurred in many parts of the Muslim world, the institutionalization of Islamic revivalism. As previously discussed, Islamic revivalism is no longer a movement of the marginalized few; it has become an institutionalized part of mainstream

Muslim life. What explains this persistent tendency to distort the nature of Islamic revivalism and to perceive it as a threat?

Secularization and Modernization

Secular presuppositions—which inform our academic disciplines and outlook on life, our Western secular worldview—have been a major obstacle to understanding Islamic politics and so have contributed to a tendency to reduce Islam to fundamentalism and fundamentalism to religious extremism. For much of the sixties the received wisdom among many, from development experts to theologians, could be summarized in the adage: "Every day in every way, things are getting more and more modern and secular." Integral to definitions of modernization were the progressive Westernization and secularization of society: its institutions, organizations and actors.[60] Religion and theology reflected the same presupposition as theologians spoke of demythologizing the scriptures, of a secular gospel for the modern age, and of the triumph of the secular city (as distinguished from Augustine's City of God), and a school of theological thought emerged which was dubbed the "Death of God Theology."[61] Religious faith was at best supposed to be a private matter. The degree of one's intellectual sophistication and objectivity in academia was often equated with a secular liberalism and relativism that seemed antithetical to religion. Although church or synagogue membership were recognized as useful, most political candidates avoided discussing their faith or religious issues.

Acceptance of the "enlightened" notion of separation of Church and State, and of Western, secular models of development, relegated religion to the stockpile of traditional beliefs, valuable in understanding the past but irrelevant or an obstacle to modern political, economic, and social development. Neither development theory nor international relations considered religion a significant variable for political analysis. The separation of religion and politics overlooked the fact that most religious traditions were established and developed in historical, political, social, and economic contexts. This was certainly true in the history of Islam and even more so in the belief of many Muslims. Ironically, some analysts became like conservative clerics the world over—they treated religious beliefs and practices as isolated, free-standing realities rather than as the product of faith and history or, more accurately, faith-in-history.

Religious traditions, while characterized as conservative or traditional, are the product of a dynamic changing process in which the Word of revelation is mediated through human interpretation or discourse in response to specific sociohistorical contexts. The post-Enlightenment tendency to define religion as a system of personal belief rather than as a way of life has seriously hampered our ability to understand the nature of Islam and many of the world's religions. It has artifically compartmentalized religion, doing violence to its nature, and reinforced a static, reified conception of religious traditions rather than revealing their inner dynamic nature. To that extent, a religion that mixes religion and politics appears retrogressive, prone to fanaticism, and thus a potential threat.

Islam has generally been regarded in the West (and among many secular-minded Muslims) as a static phenomenon doctrinally and socioculturally, and therefore antimodern and retrogressive. This attitude was supported by the prevailing tendency to emphasize such notions as the closing of the door of religious interpretation (*ijtihad*) or reform in the tenth century. Many experts were trained in area studies programs by professors (historians and social scientists) with little real expertise in Islamic religion, and in history and politics courses in which Islam was treated primarily as part of a cultural legacy, historical baggage studied for its relevance more to the past than to the present. Islamic studies themselves were textually and historically oriented with little reliance on the social sciences and with minimal attention to the modern period. Coverage of the modern period, the nineteenth and twentieth centuries, was restricted in textbooks to one chapter, often the shortest.[62] What judgments of vitality and relevance were implied by courses and analyses of modern Islam, in particular Islamic modernism, which stopped with Muhammad Abduh, who died in 1905, or Muhammad Iqbal, who died in 1938!

A second factor that has hindered our analysis of the Islamic dimension and dynamic in Muslim societies is our secular elite orientation. Focus on governments and elites reflects a bias which hampers understanding the nature of populist movements in general and Islamic movements in particular. It reinforces the notion that religion is the province of the tradition-bound cleric (mullah), religious scholars (*ulama*), and the uneducated or illiterate masses. Thus few scholars of the Middle East or other Muslim societies believed it necessary or relevant to know and meet with religious leaders, to visit their institutions, to have some idea of their leadership roles in society. Too often the *ulama* were regarded as irrelevant or of marginal interest. Most Western scholars, as well as Western-trained Muslim scholars, were more comfortable working and studying with like-minded elites in modernized, Westernized urban settings. Much of Western scholarship viewed Muslim societies through the modern prism of a development theory that was secular and

Western in its principles, values, and expectations. Experts analyzed societies as their elites ran them—from the top down. As a result, academic and government analysts and the media often slipped into the same pitfalls: focusing on a narrow, albeit powerful, secular elite segment of society; equating secularization with progress, and religion with backwardness and conservatism; and believing that modernization and Westernization are necessarily intertwined. Consequently, the options available for the political and social development of Muslim societies were often seen in polarities: the tug between tradition and modernity, the past and the future, the *madrasa* (religious college or seminary) and the university, the veil and Western dress or values.

Edward Said's critique of orientalism, though at times excessive, was insightful in identifying deficiencies and bias in the scholarship of the past.[63] However, new forms of orientalism flourish today in the hands of those who equate revivalism, fundamentalism, or the Islamic movements solely with radical revolutionaries, and who fail to focus on the vast majority of Islamically committed Muslims who belong to the moderate mainstream of society. Too often academia, government, and the media have emphasized the violent fringe and, failing to see the forest for the trees, have not studied sufficiently both moderate political and nonpolitical movements and organizations. This trend (and deficiency) has been reinforced by the realities of the marketplace. Publishing houses, journals, consulting firms, and the media seek out that which captures the headlines and all too often confirms stereotypes and fears of extremism and terrorism. Think how often any reference to an Islamic organization inevitably includes adjectives like "fundamentalist," "conservative," and "extremist."

For those who subscribe to a liberal secular or liberal Judaeo-Christian tradition, any intrusion of religion in politics is often viewed as potentially dangerous and "fundamentalist." This perception is intensified when our knowledge of religious groups is limited to those who represent a radical minority. Compare coverage of Tabligh-i Jamaat and a myriad of Jihad organizations. The former is a worldwide nonpolitical missionary Muslim organization with millions of members, while the latter are radical groups whose membership often runs only in the hundreds and whose sporadic violence, though dramatic at times, affects far fewer lives. Compare the number of substantive studies and media reports on moderate Islamic organizations such as the Egyptian Muslim Brotherhood, Pakistan's Jamaat-i-Islami, Kuwait's Jamiyyat al-Islah, or Tunisia's Renaissance Party during the past ten years with those on radical Islamic movements. As prominent as Islamic leaders and movements have been in many Muslim countries, our knowledge about their

thought, as well as about the organization and agendas of their movements, is often astonishingly limited.

The Politics of Underestimation

Why is it that with remarkable consistency the resurgence of Islam in
many countries and its political implications have often been acknowledged only at the eleventh hour? While revivalism was growing in
Egypt, Libya, Pakistan, and Malaysia in the early seventies, few noticed
or gave it any attention. In 1975 a presentation on Islam's potential role
in sociopolitical development attracted six to nine people at an International Studies Association meeting. Today at many professional meetings we have almost that number of participants on the panel alone.
What had at first been described as an epiphenomenon was often later
dismissed as a wave that had passed. This line of thought has persisted
over the past decade, as we overlooked and then discovered or were
forced to confront events in Iran, Pakistan, Lebanon, the Gulf, and now
North Africa. Many said it could happen in Pakistan but never in Iran,
or in Iran but not in Egypt, or in Egypt but most assuredly never in
Tunisia. As Islam has seemed to decline in the politics of one nation or
region, it has risen in others. Yet when the Iran–Iraq ceasefire came in
1988, many, forgetting the indigenous sources of the Islamic revival,
were quick to conclude that this signaled the discrediting of Khomeini
and thus a death blow to the "fundamentalist wave."

More often than not, Islamic movements are lumped together; conclusions are drawn, based more on stereotyping or expectations than on
empirical research. This problem is due less to the secrecy of individuals
and organizations than to more mundane factors:

   1. The dearth of scholars in Middle East Studies in general and Islamic
   Studies in particular means that most cover vast geographic areas, and
   such varied topics as politics, law, history, society, and oil.
   2. Since most scholars come with a secular academic bias and limited
   training in Islam and the Islamic dimension of Muslim societies, there is a
   greater tendency to generalize or to draw deductive conclusions from their
   secular presuppositions and prejudices.
   3. Scholars of Islam tend to be trained in and work on the past.
   4. Most experts have had limited training and interest in, and even less
   actual direct contact and extended experience with, Islamic movements
   and activists.

Yet analysis of modern Islamic movements and organizations requires more than guesswork, generalizations, and reliance solely on

newspaper reports or movement-issued documents, isolated events, or government reports. Direct observation, interaction, and study are particularly critical, since many Islamic activists write comparatively little and their writings are often mere ideological tracts or public relations documents. What they write or say must be placed within the context of what they actually do. Because of Western scholars' limited access and interest and their tendency to respond to government concerns and media interest and demands (headline events), the breadth, character, and activities of the moderate majority of contemporary organizations go relatively unnoticed.

Diversity and Change

The experience of the past decade alerts us to the need to be more attentive to the diversity behind the seeming unity of Islam, to appreciate and more effectively analyze both the unity and the diversity in Islam and in Muslim affairs. In the past, the oneness of Islam (of God, the Prophet, and the Book) gave rise to many movements and interpretations: Sunni, Shii, Kharaji, Wahhabi, and different schools of law, theology, and mysticism. So too today, differing contexts have spawned a variety of Islamically oriented nations, leaders, and organizations. The diversity of governments—the Saudi monarchy; Qaddafi's populist state; Khomeini's clerical republic in Iran; and the military regimes of Zia ul-Haq in Pakistan, and Gaafar Muhammad Nimeiri and now Omar al-Bashir in the Sudan—their differing relations with the West, and the variety of Islamic movements are undercut and distorted by the univocal connotation of the term *Islamic fundamentalism.*

Beyond the common reference to an Islamic alternative and to general ideological principles are multiple levels of discourse as one looks across the wide array of governments, societies, and Islamic organizations. There are often as many differences as similarities in Muslim interpretations of the nature of the state, Islamic law, and the status of women and minorities as there are sharp differences regarding the methods (the ballot box, social service centers, violent confrontation) to be employed for the realization of an Islamic system of government.

As a romanticized notion of early Islamic history made many Muslims forget the dynamism and degree of change in the past, so too in recent years a Western secular tendency to pit modern change against a fixed tradition has obscured the degrees of difference and change in modern Islam and modern Muslim societies. Too often, secular scholars and Muslim theologians alike have failed to appreciate the extent to which religious beliefs and laws are the product of human interpretation and application of revelation. A rigid, retrogressive Islam becomes a threat. Ironically, non-Muslim scholars sometimes sound more like mullahs. When faced with new interpretations or applications of Islam, they often critique them from the vantage point of traditional belief and practice. On the one hand, Islam is regarded as fixed, and Muslims are seen as too reluctant to accept change. On the other hand, when change occurs, it is dismissed as unorthodox, sheer opportunism, an excuse for adopting that which is outside Islam. Yet all we are witnessing is a natural process of reinterpretation by individuals and communities, another stage in the interpretation and development of the Islamic tradition. The reinterpretation of traditional Islamic concepts such as *shura* (consultation) and *ijma* (community consensus) are excellent examples of this process.

Change is a reality in contemporary Islam and in Muslim societies. It may be found at every level, in every quarter, and across social classes. The issue is not change but the amount, pace, and direction of change. The flexibility of the Islamic tradition is demonstrated not only by those whom some regard as modern reformers such as Ali Shariati or Sadiq al-Mahdi, but also by the Ayatollah Khomeini's interpretation of the doctrine of governance by the jurist (*vilayat-i-faqih*), as well as by the constitution of the Islamic Republic of Iran's acceptance of a constitutional/parliamentary form of government.

Challenge or Threat?

Is there an Islamic threat? In one sense, yes. Just as there is a Western threat or a Judaeo-Christian threat. Islam, like Christianity and Judaism, has provided a way of life which has transformed the lives of many. At the same time some Muslims, like some Christians and Jews, have also used their religion to justify aggression and warfare, conquest and persecution in the past and today. Political Islam, like the appeal to any religion or ideology, can be effective but also dangerous. Secular ideologies have proven equally vulnerable to manipulation. Spreading God's will, like spreading democracy, can become a convenient excuse for imperialism, oppression, and injustice in the name of God or the state.

Islam is also a threat to the *complacency* of Western societies—spiritually, socially, and ultimately politically. It is, in some of its forms, a straightforward questioning of both the traditions that we seem to embrace—materialism, libertinism, and individualism, though these may be only a caricature of us—and also of our commitment to the values that we say we espouse: tolerance, freedom of expression. To be challenged to "walk the way you talk" may be unsettling, but it is not in itself unfair. Likewise, political Islam is itself challenged, challenged by its own rhetoric and message to be self-critical: to live up to the standards and principles it espouses and demands of others; to avoid and denounce the excesses that are committed by governments and movements that identify themselves as Islamic; and finally, to take or share responsibility, and not simply blame the West, for the failures of Muslim societies.

The reality of Muslim societies today contributes to a climate in which the influence of Islam and activist organizations on sociopolitical development will increase rather than diminish. Muslim states continue to exist in a climate of crisis in which many of their citizens experience and speak of the failure of the state and of secular forms of nationalism and socialism. Heads of state and ruling elites or classes possess tenuous legitimacy in the face of mounting disillusionment and opposition, in the expression of which Islamic activists are often the most vocal and effective. The extent to which governments in predominantly Muslim countries fail to meet the socioeconomic needs of their societies, restrict political participation, prove insensitive to the need to effectively incorporate Islam as a component in their national identity and ideology, or appear exceedingly dependent on the West, will contribute to the appeal of an Islamic political alternative.

The political strength and durability of Islamic movements and their ideological impact are reflected in a variety of ways. They have forced government changes and, where permitted, have successfully contested elections. Rulers from Morocco to Malaysia have become more Islamically sensitive and sought to coopt religion or suppress Islamic organizations. Many have employed Islamic rhetoric and symbols more often, expanded support for Islamic mosques and schools, increased religious programming in the media, and become more attentive to public religious observances such as the fast of Ramadan or restrictions on alcohol and gambling.

When free from government repression, Islamic candidates and organizations have worked within the political system and participated in elections in Algeria, Tunisia, Turkey, Jordan, the Sudan, Egypt, Kuwait, Pakistan, and Malaysia; activists have even held cabinet-level positions in the Sudan, Pakistan, Jordan, and Malaysia. In countries such as Algeria, Tunisia, Egypt, Jordan, and Pakistan, Islamic organizations have been among the best-organized opposition forces, and are often willing to form alliances or cooperate with political parties, professional syndicates, and voluntary associations to achieve shared political and socioeconomic re-

forms. Islamic student organizations successfully compete in student elections in the universities and lead student strikes and demonstrations.

While the vast majority of Islamic organizations are moderate and
work within the system, clandestine radical organizations that advocate the violent seizure of power to establish an Islamic state continue
to exist. A minority of militant Islamic organizations with names like
al-Jihad, the Party of God, Salvation from Hell, the Army of God,
and the Islamic Liberation Organization will continue to resort to violence and terrorism.

If the Iranian Revolution dominated much of Middle Eastern politics
and Western perceptions of the Muslim world during the eighties, the
nineties will prove to be a decade of new alliances and alignments in
which Islamic movements will challenge rather than threaten their societies and the West. Contemporary Islam challenges the West to know
and understand the diversity of the Muslim experience. It challenges
Muslim governments to be more responsive to popular demands for
political liberalization and greater popular participation, to tolerate
rather than repress opposition movements and build viable democratic
institutions. At the same time it challenges Western powers to stand by
the democratic values they embody and recognize authentic populist
movements and the right of the people to determine the nature of their
governments and leadership.

American support for repressive regimes will intensify anti-Americanism, as events in the Shah's Iran, Lebanon, Nimeiri's Sudan, and the
West Bank and Gaza have demonstrated. One can neither deny nor
overlook the fact that there is often a strong anti-Western, and especially
anti-American, sentiment among many moderates as well as radicals,
secular as well as Islamic. It is manifested in a tendency to regard the
United States as anti-Islamic and uncritically pro-Israeli, and to blame
the ills of Muslim societies upon Western political, economic, and sociocultural influences. At the same time the double equation of "Kho-
meinism" with "Islam" and of "violent radicalism" with "fundamentalism" results in the assumption that Islamic movements are naturally or
inherently anti-Western, thus obscuring the causes of anti-Americanism
and radicalism. Movements are more often motivated by objection to
specific Western policies than by cultural hostility. Differences between
Western and Muslim societies can best be explained by competing political, socioeconomic, and cultural interests. U.S. presence and policy, not
a genetic hatred for Americans, is often the primary motivating force
behind acts against American government, business, and military interests. American interests will best be served by policies that walk the fine

line between selective, discreet, low-visibility cooperation with friendly Muslim governments, and a clear, consistent public policy concerning the rights of citizens to determine their future democratically.

The assumption that the mixing of religion and politics necessarily and inevitably leads to fanaticism and extremism has been a major factor in our concluding that Islam and democracy are incompatible. Failure to differentiate between Islamic movements—between those that are moderate and those that are violent and extremist—is simplistic and counterproductive. The American government does not equate the actions of Jewish or Christian extremist leaders or groups with Judaism and Christianity as a whole. Similarly, the American government does not condemn the mixing of religion and politics in Israel, Poland, Eastern Europe, or Latin America. A comparable level of discrimination is absent when dealing with Islam. President Reagan's linking of Qaddafi and Libyan terrorism with a worldwide Muslim fundamentalist movement in his announcement of the U.S. bombing of Libya confirmed what many saw as America's monolithic, anti-Islamic approach to the Muslim world. Vice President Quayle's address to the graduating class of Annapolis in 1990, linking Nazism, communism, and radical Islamic fundamentalism, demonstrated a similar ill-informed position.

U.S. perception of a monolithic "Islamic threat" often contributes to support for repressive governments in the Muslim world and thus to the creation of a self-fulfilling prophecy. The thwarting of a participatory political process by governments that cancel elections or repress populist Islamic movements that prove effective in electoral politics, as in Tunisia and Algeria, encourages radicalization. Violence begets violence. Many of those who experience regime violence (harassment, imprisonment, torture) or see their colleagues languish and die in prison will conclude that seeking democracy is a dead end. They will withdraw from the political process and become convinced that violence is their only recourse. U.S. official silence or economic and political support for regimes is read as complicity and a sign of America's double standard for the implementation of democracy. Furthermore, it contributes to the creation of "self-fulfilling prophecies." Government repression and violation of human rights and a compliant U.S. policy toward such actions can lead to political confrontation and violence, enabling Muslim governments and some U.S. policymakers to seemingly validate their prior contention that Islamic movements are inherently violent, antidemocratic, and a threat to national and regional stability.

Contrary to what some have advised, the United States should not in principle object to the implementation of Islamic law or the involvement of Islamic activists in government. Islamically oriented political actors

and groups should be evaluated by the same criteria as any other potential political leaders or opposition parties. While some are rejectionists,
most Islamically oriented leaders or governments will be critical and
selective in their relations with the United States. However, they will
generally operate on the basis of national interests and demonstrate a
flexibility that reflects acceptance of the realities of a globally interdependent world. The United States must be willing to demonstrate by word
and action its belief that the right of self-determination and representative government includes acceptance of an Islamically oriented state and
society, if that reflects the popular will and does not directly threaten
U.S. interests.

The United States should avoid being seen as intervening in state-
initiated Islamization programs, or as opposing the activities of Islamic
organizations, where such programs or activities pose no threat to it.
American policy should, in short, be carried on in the context in which
ideological differences between the West and Islam are recognized and,
to the greatest extent possible, accepted or at least tolerated.

For more than a decade "Islamic fundamentalism" has been increasingly
identified as a threat by and to governments in the Muslim world and the
West. This belief has been informed by the impact of the Iranian Revolution, the specter of its export, the identification of Qaddafi and Khomeini with worldwide terrorism, the image of Anwar Sadat slain by
Islamic extremists, and the denunciation of the West and attacks against
Western installations and personnel by shadowy groups in Lebanon and
elsewhere. Many watched and warned of other Irans or of radical groups
seizing power through political assassination. Others soon dismissed Islamic revivalism as an epiphenomenon, only to rediscover it as a threat
when it reappeared in a new context. The many faces of contemporary
Islamic revivalism tended to be subsumed under the monolith of "Islamic fundamentalism," which was equated with violence and fanaticism, with mullah-led theocracies or small, radical guerrilla groups.

The nineties have challenged these presumptions and expectations.
There have been no other Iranian-style Islamic revolutions, nor have
any radical groups seized power. Yet neither the end of the Iran–Iraq
war nor the death of the Ayatollah Khomeini signaled the passing of
Islamic revivalism. The resurgence of Islam in Muslim politics has been
far more indigenously rooted. It has not receded but rooted itself more
deeply and pervasively. Its many faces and postures—long overshadowed by the equation of Islamic revivalism with radical fundamentalism
or with the Ayatollah Khomeini—has surfaced and will continue to do
so. The impact of revivalism can now be seen by the extent to which it

has become part of mainstream Muslim life and society and not simply the province of marginalized and alienated groups. Ironically, this has led many to see it as an even greater threat. Secular institutions are now complemented or challenged by Islamically oriented schools, clinics, hospitals, banks, publishing houses, and social services. Their ability to provide much-needed services is often taken by regimes as an implicit if not explicit critique or threat, underscoring the regimes' limitations and failures. Similarly, the emergence of an alternative elite, modern-educated but more Islamically oriented, challenges the Western, secular presuppositions and life-styles of many in the establishment.

More ominously for some, many Islamic movements in recent years have joined the rising chorus of voices calling for political liberalization. From North Africa to Southeast Asia they have participated in electoral politics and, relative to the expectations of some, scored stunning successes. This has created a political and analytical dilemma. The justification for the condemnation and suppression of Islamic movements was that they were violent extremists—small, nonrepresentative groups on the margins of society who refused to work within the system and as such were a threat to society and regional stability. The vision of Islamic organizations working today within the system has ironically made them an even more formidable threat to regimes in the Muslim world and to some in the West. Those who once dismissed their claims as unrepresentative and who denounced their radicalism as a threat to the system now accuse them of an attempt to "hijack democracy."

Islamic movements are indeed a challenge to the established order of things, to the presuppositions that have guided many governments and policymakers. The tendency to focus on the more Western-oriented and secular elite minority, and to transform secular predispositions into guidelines for sociopolitical development, blinded many to deeper social realities. In many Muslim societies religion remains a pervasive, though at times diffuse, social force, and popular political culture is far less secular than is often presumed. The power of an idea or belief, when coupled with the economic and political failures of established governments, was neither anticipated nor comprehensible for those more accustomed to secular isms—nationalism, socialism, communism. As a result, the shock of the Iranian Revolution and more recently the electoral strength of Islamic movements in Algeria, Tunisia, Egypt, Jordan, and the Sudan forced many to confront the unthinkable. This challenge in the name of Islam to the conventional secular worldview which has constituted our norm, is often dismissed as deviant, irrational, extremist. For liberal, secularly informed Western intellectuals, policymakers, and experts as well as many elites in the Muslim world, religion in public life necessarily constitutes a retrogressive fundamentalist threat, whether it be Muslim or Christian. For many governments in the Muslim world, whose legitimacy is tenuous and whose power is based upon coercion, the combination of "uncontrolled democracy" and Islam is indeed a formidable threat. For Western governments, long accustomed to pragmatic alliances with regimes which, however undemocratic or repressive, were dominated by Western-oriented elites, the leap into the unknown of a potential fundamentalist government is far from attractive. As a result, the challenge of contemporary Islamic revivalism to the political and intellectual establishment is easily transformed into a threat.

The challenge need not always result in a threat to regional stability or Western interests. Certainly those groups, whether secular or Islamic, who attempt to impose their will through assassination and violent revolution are a threat. Furthermore, many contemporary Islamic movements often appear to challenge the very principles of self-determination and intellectual and political pluralism that we espouse. Therefore populist movements that participate in political life constitute a twofold challenge. On the one hand, governments in the Muslim world and in the West that espouse political liberalization and democracy are challenged to remain true to these very principles. On the other hand, Islamic movements, should they come to power, will be challenged to extend to their opposition and to minorities the very principles of political pluralism and participation which they now demand for themselves. All are challenged to recognize that democratization is a process which entails experimentation and is necessarily accompanied by success and failure. The transformation of the West from feudal monarchies to democratic nation-states took time, trial, and error. It was accompanied by political as well as intellectual revolutions which rocked both State and Church. The transformation of political culture, values, and institutions does not happen overnight. It is a long-drawn-out process marked by debate and battle among factions with competing visions and interests.

Today we are witnessing a new historical transformation. Countries in the Muslim world, as in the former Soviet Union, Eastern Europe, and other parts of the world, are undergoing a process of political liberalization or democratization that was prevented in the past by colonialism and more recently by authoritarian governments. Risks exist, for there can be no risk-free democracy. Those who fear the unknown—what specific Islamic movements in power will be like or how they will act—have a legitimate concern. However, in principle if we condemn government repression of the opposition, cancellation of elections, and massive violation of human rights, then the same concern must apply to many current regimes such as those in Tunisia and Algeria, where the results of the Arab world's first free parliamentary elections have been denied and the winners imprisoned. The track record of these governments on political participation and pluralism and respect for human rights we know already.[64]

As some dream of the creation of a New World Order, and many millions in North Africa, the Middle East, Central Asia, and southern and Southeast Asia aspire to greater political liberalization and democratization, the continued vitality of Islam and Islamic movements need not be a threat but a challenge. For many Muslims, Islamic revivalism is a social rather than a political movement whose goal is a more Islamically minded and oriented society, but not necessarily the creation of an Islamic state. For others, the establishment of an Islamic order requires the creation of an Islamic state. In either case, Islam and most Islamic movements are not necessarily anti-Western, anti-American, or anti-democratic. While they are a challenge to the outdated assumptions of the established order and to autocratic regimes, they do not necessarily threaten American interests. Our challenge is to better understand the history and realities of the Muslim world. Recognizing the diversity and many faces of Islam counters our image of a unified Islamic threat. It lessens the risk of creating self-fulfilling prophecies about the battle of the West against a radical Islam. Guided by our stated ideals and goals of freedom and self-determination, the West has an ideal vantage point for appreciating the aspirations of many in the Muslim world as they seek to define new paths for their future.

NOTES

Introduction

1. Daniel Pipes, "Fundamentalist Muslims," *Foreign Affairs,* Summer 1986 pp. 939–59.
2. Strobe Talbott, "Living with Saddam," *Time,* February 25, 1991, p. 24.
3. Patrick J. Buchanan, "Is Islam an Enemy of the United States?," *New Hampshire Sunday News,* November 25, 1990.
4. "Islam and the West," *The Economist,* December 22, 1990, p. 18.

Chapter 1

1. John O. Voll, "Renewal and Reform in Islamic History: *Tajdid* and *Islah,*" in John L. Esposito, ed., *Voices of Resurgent Islam* (New York: Oxford University Press, 1983); Nehemiah Levtzion and John O. Voll, eds., *Eighteenth-Century Renewal and Reform in Islam* (Syracuse, N.Y.: Syracuse University Press, 1987).
2. Donald Eugene Smith, ed., *Religion and Modernization* (New Haven, Conn.: Yale University Press, 1974), p. 4.
3. Daniel Crecelius, "The Course of Secularization in Modern Egypt," in John L. Esposito, ed., *Islam and Development: Religion and Sociopolitical Change* (Syracuse, N.Y.: Syracuse University Press, 1980), p. 60.
4. For studies of the Islamic resurgence see Yvonne Y. Haddad, John O. Voll, and John L. Esposito, eds., *The Contemporary Islamic Revival: A Critical Survey and Bibliography* (New York: Greenwood Press, 1991); John L. Esposito, *Islam and Politics,* 3rd ed. (Syracuse, N.Y.: Syracuse University Press, 1991); John L. Esposito, ed., *Islam in Asia: Religion, Politics and Society* (New York: Oxford University Press, 1987); James P. Piscatori, ed., *Islam in the Political Process* (Cambridge: Cambridge University Press, 1983); and Nazih Ayubi, *Political Islam: Religion and Politics in the Arab World* (London: Routledge, 1991).
5. See John J. Donohue, "Islam and the Search for Identity in the Arab

World," in Esposito, ed., *Voices of Resurgent Islam,* ch. 3; Ali Merad, "The Ideologisation of Islam in the Contemporary Arab World," in Alexander S. Cudsi and Ali E. Hillal Dessouki, eds., *Islam and Power* (Baltimore, Md.: The Johns Hopkins University Press, 1981), ch. 3; Yvonne Y. Haddad, "The Arab–Israeli Wars, Nasserism, and the Affirmation of Islamic Identity," in John L. Esposito, ed., *Islam and Development: Religion and Sociopolitical Change* (Syracuse, N.Y.: Syracuse University Press, 1980), pp. 118–20.

6. Philip S. Khoury, "Islamic Revivalism and the Crisis of the Secular State in the Arab World," in I. Ibrahim, ed., *Arab Resources: The Transformation of a Society* (London: Croom Helm, 1983), pp. 213–34; Ayubi, *Political Islam,* pp. 164–74; Henry Munson, *Islam and Revolution* (New Haven, Conn.: Yale University Press, 1988), ch. 9.

7. James P. Piscatori, *Islam in a World of Nation States* (Cambridge: Cambridge University Press, 1986), p. 26.

8. The Islamic (Student) Association of Cairo University, "Lessons from Iran," in John J. Donohue and John L. Esposito, eds., *Islam in Transition: Muslim Perspectives* (New York: Oxford University Press, 1982), p. 246.

9. Ibid., p. 247.

10. See Michael C. Hudson, "The Islamic Factor in Syrian and Iraqi Politics," in James P. Piscatori, ed., *Islam in the Political Process* (Cambridge: Cambridge University Press, 1983), pp. 86ff.

11. Jacob Goldberg, "The Shii Minority in Saudi Arabia," in Juan R. I. Cole and Nikki R. Keddie, eds., *Shiism and Social Protest* (New Haven, Conn.: Yale University Press, 1986), p. 243.

12. William Ochsenwald, "Saudi Arabia and the Islamic Revival," *International Journal of Middle East Studies* 13:3 (August 1981): 271, 276–77; Joseph Kechichian, "The Role of the Ulama in the Politics of an Islamic State: The Case of Saudi Arabia," *International Journal of Middle East Studies* 18:1 (February 1986); 56–68; James P. Piscatori, "Ideological Politics in Saudi Arabia," in Piscatori, *Islam in the Political Process* (Cambridge: Cambridge University Press, 1983), p. 67.

13. Shaul Bakhash, *The Reign of the Ayatollahs: Iran and the Islamic Revolution* (New York: Basic Books, 1984), pp. 234–35.

14. Delip Hiro, *Iran Under the Ayatollahs* (London: Routledge and Kegan Paul, 1985), p. 340.

15. Joseph Kostiner, "Kuwait and Bahrain," in Shireen T. Hunter, ed., *The Politics of Islamic Revivalism* (Bloomington: Indiana University Press, 1988), pp. 121ff.

Chapter 2

1. For a Western scholar's perspective, see Francis E. Peters, *Children of Abraham* (Princeton, N.J.: Princeton University Press, 1982). To see the basic texts of these three religious systems juxtaposed on common issues, see Peters's *Judaism, Christianity, and Islam: The Classical Texts and Their Interpretation,* 3 vols. (Princeton, N.J.: Princeton University Press, 1990).

2. For an introduction to Islam from the formation of the early Muslim community to the present, see John L. Esposito, *Islam: The Straight Path,* expanded ed. (New York: Oxford University Press, 1991); Frederick Mathewson Denny, *An Introduction to Islam* (New York: Macmillan, 1985); Fazlur Rahman, *Islam,* 2nd ed. (Chicago: University of Chicago Press, 1979).

3. For a translation of this classic, see A. Guillaume, *The Life of Muhammad: A Translation of Ishaq's Sirat Rasul Allah* (London: Oxford University Press, 1955).

4. The standard works in English remain W. Montgomery Watt, *Muhammad at Mecca* (Oxford: Clarendon Press, 1953), and *Muhammad at Medina* (Oxford: Clarendon Press, 1956). The two volumes were subsequently published in an abridged version, *Muhammad: Prophet and Statesman* (London: Oxford University Press, 1961).

5. Fred McGraw Donner, *The Early Islamic Conquests* (Princeton, N.J.: Princeton University Press, 1981), pp. 269ff.

6. Hugh Kennedy, *The Prophet and the Age of the Caliphates* (London: Longman, 1986).

7. Donner, *Early Islamic Conquests,* p. 269.

8. Bernard G. Weiss and Arnold H. Green, *A Survey of Arab History* (Cairo: The American University of Cairo Press, 1987), p. 59.

9. For the history of this period, see Ira Lapidus, *A History of Islamic Societies* (Cambridge: Cambridge University Press, 1988).

10. Patrick J. Bannerman, *Islam in Perspective* (London: Routledge, 1988), p. 86.

11. Marshall G. S. Hodgson, *The Venture of Islam,* 3 vols. (Chicago: University of Chicago Press, 1974), 1:235.

12. The standard works on Islamic law are Noel J. Coulson, *A History of Islamic Law* (Edinburgh: Edinburgh University Press, 1964), and Joseph Schacht, *An Introduction to Islamic Law* (Oxford: Clarendon Press, 1964).

13. For insightful introductions to Sufism, see Annemarie Schimmel, *The Mystical Dimensions of Islam* (Chapel Hill: University of North Carolina Press, 1975); A. J. Arberry, *An Introduction to the History of Sufism* (London: Longman, 1942); Martin Lings, *What Is Sufism?* (Berkeley: University of California Press, 1977).

14. Maxime Rodinson, "The Western Image and Western Studies of Islam," in Joseph Schacht and C. E. Bosworth, eds., *The Legacy of Islam* (Oxford: Oxford University Press, 1974), p. 9.

15. Albert Hourani, *Europe and the Middle East* (Berkeley: University of California Press, 1980), p. 9.

16. R. Stephen Humphreys, *Islamic History: A Framework for Inquiry* (Minneapolis, Minn.: Bibliotheca Islamica, 1988), p. 250.

17. Francis E. Peters, "The Early Muslim Empires: Umayyads, Abbasids, Fatimids," in Marjorie Kelly, ed., *Islam: The Religious and Political Life of a World Community* (New York: Praeger, 1984), p. 79.

18. For a history of the Crusades, see S. Runciman, *A History of the Crusades,* 3 vols. (Cambridge: Cambridge University Press, 1951–54), and J. Prawers, *Crusader Institutions* (Oxford: Oxford University Press, 1980).

19. J. J. Saunders, *A History of Medieval Islam* (London: Routledge and Kegan Paul, 1965), p. 154.

20. Peters, "Early Muslim Empires," p. 85.

21. Saunders, *History,* p. 161.

22. C. E. Bosworth, "The Historical Background of Islamic Civilization," in R. M. Savory, ed., *Introduction to Islamic Civilization* (New York: Cambridge University Press, 1980), p. 25.

23. Ira Lapidus, *A History of Islamic Societies* (New York: Cambridge University Press, 1988), p. 330.

24. Arthur Goldschmidt, Jr., *A Concise History of the Middle East,* 3rd ed. (Boulder, Colo.: Westview Press, 1988), p. 132.

25. Paul Coles, *The Ottoman Impact on Europe* (New York: Harcourt, Brace and World, 1968), pp. 146–47.

26. Ibid., p. 147.

27. Ibid., p. 151.

28. Ibid., p. 148.

29. Bosworth, "Historical Background," p. 25.

30. Ibid., p. 139.

31. Hourani, *Europe and the Middle East,* p. 10.

32. R. W. Southern, *Western Views of Islam and the Middle Ages* (Cambridge, Mass.: Harvard University Press, 1962), p. 2.

33. Ibid., p. 28.

34. Ibid., p. 31.

35. Rodinson, "Western Image," p. 14.

36. Hourani, *Europe and the Middle East,* p. 10.

37. Ibid., p. 12.

Chapter 3

1. Hichem Djait, *Europe and Islam: Cultures and Modernity* (Berkeley: University of California Press, 1985), p. 148.

2. Albert Hourani, *Europe and the Middle East* (Berkeley: University of California Press, 1980), pp. 12–13.

3. Maxime Rodinson, *Europe and the Mystique of Islam* (Seattle: University of Washington Press, 1987), p. 66.

4. John O. Voll, "Renewal and Reform in Islamic History: *Tajdid* and *Islah,*" in John L. Esposito, ed., *Voices of Resurgent Islam* (New York: Oxford University Press, 1983), ch. 2.

5. John O. Voll, *Islam: Continuity and Change in the Modern World* (Boulder, Colo.: Westview Press, 1982), ch. 3.

6. Ibid., p. 33.

7. Wilfred Cantwell Smith, *Islam in Modern History* (Princeton, N.J.: Princeton University Press, 1957), p. 41.

8. As quoted in Asma Rashid, "A Critical Appraisal of James J. Cooke's 'Tricolour and Crescent: Franco-Muslim Relations in Colonial Algeria, 1880–1940,' " *Islamic Studies* 29:2 (Summer 1990): 203.

9. Arthur Goldschmidt, Jr., *A Concise History of the Middle East,* 3rd ed. (Boulder, Colo.: Westview Press, 1988), p. 231.

10. Ibid., p. 204.

11. Ibid., pp. 204–5.

12. William Wilson Hunter, *Indian Musulmans* (London, 1871), p. 184.

13. Hafeez Malik, *Sir Sayyid Ahmad Khan and Muslim Modernization in India and Pakistan* (New York: Columbia University Press, 1980), p. 172.

14. Albert Hourani, *Arabic Thought in the Liberal Age* (Oxford: Oxford University Press, 1970), especially chs. 2–7.

15. John O. Voll, "Islamic Renewal and the Failure of the West," in Richard T. Antoun and Mary Elaine Hegland, eds., *Religious Resurgence: Contemporary Cases in Islam, Christianity, and Judaism* (Syracuse, N.Y.: Syracuse University Press, 1987), ch. 6.

16. Hourani, *Arabic Thought,* p. 109.

17. Ibid., p. 122.

18. Muhammad Iqbal, *The Reconstruction of Religious Thought in Islam* (Lahore: Muhammad Ashraf, 1968), p. 8.

19. Ibid., p. 163.

20. Muhammad Iqbal, "Islam as a Social and Political Ideal," in S. A. Vahid, ed., *Thoughts and Reflections of Iqbal* (Lahore: Muhammad Ashraf, 1964), p. 35.

21. Taha Husayn, "The Future of Culture in Egypt," in John J. Donohue and John L. Esposito, eds., *Islam in Transition: Muslim Perspectives* (New York: Oxford University Press, 1982), pp. 74–75.

22. Ibid., p. 74.

23. Ibid., p. 75.

24. Albert Hourani, *A History of the Arab Peoples* (Cambridge, Mass.: Harvard University Press, 1990), pp. 308ff.

25. S. Abul Hasan Ali Nadwi, *Islam and the World* (n.d.; rpt. Lahore: Muhammad Ashraf, 1967), p. 139.

26. Abul Ala Mawdudi, *Nationalism and Islam* (Lahore: Islamic Publications, 1947), p. 10.

27. Ira M. Lapidus, *A History of Islamic Societies* (Cambridge: Cambridge University Press, 1988), pp. 625–26.

28. Richard P. Mitchell, *The Society of the Muslim Brothers* (London: Oxford University Press, 1969), p. 229.

29. Goldschmidt, *Concise History,* p. 278.

30. Ali E. Hillal Dessouki, "The Limits of Instrumentalism: Islam in Egypt's Foreign Policy," in Adeed Dawisha, ed., *Islam in Foreign Policy* (Cambridge: Cambridge University Press, 1984), p. 87.

Chapter 4

1. Quoted in Mahmoud Mustafa Ayoub, *Islam and the Third Universal Theory: The Religious Thought of Muammar al-Qadhdhafi* (London: KPI, 1987), p. 32. For a more extended analysis of the role of Islam in politics, see

John L. Esposito, *Islam and Politics,* 3rd ed. (Syracuse, N.Y.: Syracuse University Press, 1991), which is a source for this discussion.

2. Raymond N. Habiby, "Muamar Qadhafi's New Islamic Scientific Socialist Society," in Michael Curtis, ed., *Religion and Politics in the Middle East* (Boulder, Colo.: Westview Press, 1981), p. 247.

3. Lillian Craig Harris, *Libya: Qadhafi's Revolution and the Modern State* (Boulder, Colo.: Westview Press, 1986), p. 31.

4. Lisa Anderson, "Qaddafi's Islam," in John L. Esposito, ed., *Voices of Resurgent Islam* (New York: Oxford University Press, 1983), ch. 6, and Ann Elizabeth Mayer, "Islamic Resurgence or New Prophethood: The Role of Islam in Qaddafi's Ideology," in Ali E. Hillal Dessouki, ed., *Islamic Resurgence in the Arab World* (New York: Praeger, 1982), ch. 10.

5. "The Libyan Revolution in the Words of Its Leaders," *Middle East Journal* 24 (1970): 208.

6. Muammar al-Qdhadhafi, *The Green Book,* excerpted as the "Third Way," in John J. Donohue and John L. Esposito, eds., *Islam in Transition: Muslim Perspectives* (New York: Oxford University Press, 1982), p. 103.

7. Oriana Fallaci, "The Iranians Are Our Brothers: An Interview with Colonel Muammar al-Qaddafi of Libya," *New York Times Magazine,* December 16, 1979.

8. Ayoub, *Islam,* p. 94.

9. Dirk Vandewalle, "Qadhafi's Unfinished Revolution," *Mediterranean Quarterly,* Winter 1990, p. 72.

10. Lisa Anderson, "Tunisia and Libya: Responses to the Islamic Impulse," in John L. Esposito, ed., *The Iranian Revolution: Its Global Impact* (Miami: Florida International University Press, 1990), p. 171.

11. George Joffe, "Islamic Opposition in Libya," *Third World Quarterly* 10:2 (April 1988): 624.

12. Anderson, "Qaddafi's Islam," p. 143.

13. Ibid., p. 629.

14. Jennifer Parmelee, "At Home, Gadhafi May Face Religious Turmoil," *International Herald Tribune,* January 11, 1989.

15. Adel Darwish, "Gaddafi's Garbled Message," *The Middle East,* December 1989, p. 14.

16. Vandewalle, "Qadhafi's Unfinished Revolution," p. 80.

17. B. Scarcia Amoretti, "Libyan Loneliness in Facing the World: The Challenge of Islam," in Adeed Dawisha, ed., *Islam in Foreign Policy* (Cambridge: Cambridge University Press, 1983), pp. 55ff.

18. Sulayman S. Nyang, "The Islamic Factor in Libya's Foreign Policy," *Africa and the World,* 1, no. 2 (1987), p. 20.

19. Lela Garner Noble, "The Philippines: Autonomy for Muslims," in John L. Esposito, ed., *Islam in Asia: Religion, Politics & Society* (New York: Oxford University Press, 1987), p. 101.

20. Nyang, "Islamic Factor," pp. 17–18.

21. *Al-Nahj al-Islami limadha* (Cairo: al-Maktab al Misri al-Hadith, 1980).

22. *The Arab News,* May 31, 1984.

23. Peter Bechtold, "The Sudan Since the Fall of Numayri," in Robert O. Freedman, ed., *The Middle East from the Iran-Contra-Affair to the Intifada* (Syracuse, N.Y.: Syracuse University Press, 1990), p. 371.

24. John O. Voll, "Political Crisis in the Sudan," *Current History* 89:546 (April 1990): 179.

25. Ann Lesch, "Khartoum Diary," *Middle East Report,* no. 161 (November–December, 1989), p. 37.

26. Voll, "Political Crisis," pp. 179–80.

27. Dessouki, *Islamic Resurgence,* p. 90.

28. Amira El-Azhary Sonbol, "Egypt," in Shireen T. Hunter, ed., *The Politics of Islamic Revivalism* (Bloomington: Indiana University Press, 1988), p. 25.

29. Gilles Kepel, *Muslim Extremism in Egypt: The Prophet and Pharaoh* (Berkeley: University of California Press, 1986), p. 192.

30. Johannes J. G. Jansen, *The Neglected Duty* (New York: Macmillan, 1986), p. 193.

31. Dessouki, *Islamic Resurgence,* p. 91.

32. John Kifner, "Egyptian Opposition Lionizes Guard Who Killed 7 Israelites," *New York Times,* December 27, 1985.

33. Jane Freedman, "Democratic Winds Blow in Cairo," *Christian Science Monitor,* January 17, 1990.

34. Ervand Abrahamian, *Iran: Between Two Revolutions* (Princeton, N.J.: Princeton University Press, 1982), p. 448.

35. James A. Bill, *The Eagle and the Lion* (New Haven, Conn.: Yale University Press, 1988), p. 176.

36. In order of quotation, ibid., pp. 186, 184.

37. Ibid., p. 186.

38. As quoted in Ervand Abrahamian, "Ali Shariati: Ideologue of the Iranian Revolution," *MERIP Reports* 102 (January 1982): 26.

39. Dr. Ali Shariati, *Man and Islam* (Houston, Tex.: Free Islamic Literature, 1980), p. xi.

40. As quoted in Ali Shariati, *On the Sociology of Islam* (Berkeley, Calif.: Mizan Press, 1979), p. 23.

41. As quoted in Abrahamian, "Ali Shariati," p. 26.

42. Donohue and Esposito, eds., *Islam in Transition,* p. 302.

43. Ruhollah Khomeini, *Islam and Revolution: Writings and Declarations of Imam Khomeini* (Berkeley, Calif.: Mizan Press, 1981), pp. 249–50.

44. For an analysis of Khomeini's worldview, see Farhang Rajaee, *Islamic Values and the World: Khomeini on Man, the State and International Politics* (Washington, D.C.: University Press of America, 1983).

45. "Message to the Pilgrims," in *Islam and Revolution: Writings and Declarations of Imam Khomeini,* trans. Hamid Algar (Berkeley, Calif.: Mizan Press, 1981), p. 195.

46. Said Amir Arjomand, *The Turban for the Crown: The Islamic Revolution in Iran* (New York: Oxford University Press, 1987), p. 164.

47. Ibid.

48. For an analysis of the global impact of the Iranian Revolution and its implications for U.S. foreign policy, see John L. Esposito, ed., *The Iranian Revolution: Its Global Impact* (Miami: Florida International University Press, 1990).

49. John L. Esposito and James P. Piscatori, "The Global Impact of the Iranian Revolution: A Policy Perspective," ibid., p. 317.

50. "Constitution of the Islamic Republic of Iran," *Middle East Journal* 34 (1980): 185.

51. Ayatollah Khomeini as quoted in Esposito and Piscatori, "Global Impact," p. 322.

Chapter 5

1. Robin Wright, *Sacred Rage: The Wrath of Militant Islam* (New York: Simon and Schuster, 1985).

2. Richard Mitchell, *The Society of Muslim Brothers* (New York: Oxford University Press, 1969), and Charles J. Adams, "Mawdudi and the Islamic State," in John L. Esposito, ed., *Voices of Resurgent Islam* (New York: Oxford University Press, 1983), ch. 5.

3. John L. Esposito, *Islam and Politics,* 3rd ed. (Syracuse, N.Y.: Syracuse University Press, 1991), pp. 130–50.

4. Hassan al-Banna, "The New Renaissance," in John J. Donohue and John L. Esposito, eds., *Islam in Transition: Muslim Perspectives* (New York: Oxford University Press, 1982), p. 78.

5. Abul Ala Mawdudi, "Political Theory of Islam," in Donohue and Esposito, eds., *Islam in Transition,* p. 252.

6. Mitchell, *Society,* pp. 228–29.

7. Hassan al-Banna, "The New Renaissance," p. 78.

8. Muhammad Yusuf, *Maududi: A Formative Phase* (Karachi: The Universal Message, 1979), p. 35.

9. Sayyid Qutb, "Social Justice in Islam," in Donohue and Esposito, eds., *Islam in Transition,* pp. 125–26.

10. Yvonne Y. Haddad, "Sayyid Qutb: Ideologue of Islamic Revival," in Esposito, ed., *Voices of Resurgent Islam,* p. 69.

11. Adnan Ayyuh Musallam, "The Formative Stages of Sayyid Qutb's Intellectual Career and His Emergence as an Islamic Da'iyah, 1906–1952," Ph.D. diss., University of Michigan, 1983, p. 225. See also Gilles Kepel, *Muslim Extremism in Egypt: The Prophet and Pharaoh* (London: Al-Saqi Books, 1985), p. 41.

12. Yvonne Y. Haddad, "Sayyid Qutb," pp. 77ff.

13. Kepel, *Muslim Extremism in Egypt;* Emmanuel Sivan, *Radical Islam: Medieval Theology and Modern Politics* (New Haven, Conn.: Yale University Press, 1985); and Johannes Jansen, *The Neglected Duty: The Creed of Sadat's Assassins and Islamic Resurgence in the Middle East* (New York: Macmillan, 1986).

14. Kepel, *Muslim Extremism in Egypt,* pp. 31–32.

15. Saad Eddin Ibrahim, "Egypt's Islamic Activism in the 1980's," *Third World Quarterly,* April 1988, p. 643.

16. Saad Eddin Ibrahim, "Islamic Militancy as a Social Movement: The Case of Two Groups in Egypt," in Ali E. Hillal Dessouki, ed., *Islamic Resurgence in the Arab World* (New York: Praeger, 1982), p. 118.

17. Hamied N. Ansari, "The Islamic Militants in Egyptian Politics," *International Journal of Middle East Studies* 16:1 (March 1984): 123–44.

18. Johannes J. G. Jansen, *The Neglected Duty* (New York: Macmillan, 1986), p. 169.

19. Ibid., p. 161.

20. Ibid., p. 193.

21. *New York Times,* September 3, 1989.

22. Fouad Ajami, "In the Pharaoh's Shadow: Religion and Authority in Egypt," in James P. Piscatori, ed., *Islam in the Political Process* (Cambridge: Cambridge University Press, 1983), p. 29.

23. Ibid., p. 134.

24. Saad Eddin Ibrahim, "Egypt's Islamic Militants," in *MERIP Reports* 103 (February 1982): 11. See also Saad Eddin Ibrahim, "Islamic Militancy as a Social Movement," in Dessouki, ed., *Islamic Resurgence,* pp. 128–31, and Sivan, *Radical Islam,* pp. 118–19.

25. Louis J. Cantori, "The Islamic Revival as Conservatism and as Progress in Contemporary Egypt," in Emile Sahliyeh, ed., *Religious Resurgence and Politics in the Contemporary World* (Albany: SUNY Press, 1990), p. 191.

26. See Michael Hudson, *The Precarious Republic* (New York: Random House, 1968).

27. Augustus Richard Norton, "Religious Resurgence and Political Mobilization of the Shia in Lebanon," in Sahliyeh, ed., *Religious Resurgence,* p. 231.

28. Ralph E. Crow, "Religious Sectarianism in the Lebanese Political System," *Journal of Politics* 24 (August 1962): 489–520; and Albert Hourani, "Lebanon: The Development of a Political System," in Leonard Binder, ed., *Politics in Lebanon* (New York: John Wiley, 1966), pp. 13–29.

29. Joseph Olmert, "The Shiis and the Lebanese State," in Martin Kramer, ed., *Shiism, Resistance and Revolution* (Boulder, Colo.: Westview Press, 1987), p. 197.

30. Robin Wright, "Lebanon," in Shireen T. Hunter, ed., *The Politics of Islamic Revivalism* (Bloomington: Indiana University Press, 1988), pp. 62–63.

31. R. Augustus Norton, "Harakat AMAL—The Emergence of a New Lebanon Fantasy or Reality," in *Islamic Fundamentalism and Islamic Radicalism* (Washington, D.C.: U.S. Government Printing Office, 1985), p. 347.

32. Fouad Ajami, *The Vanished Imam* (Ithaca, N.Y.: Cornell University Press, 1986), p. 199.

33. Ibid., pp. 189–90.

34. Robin Wright, "Lebanon," in Shireen T. Hunter, ed., *The Politics of Islamic Revivalism* (Bloomington: Indiana University Press, 1988), p. 66.

35. Experts differ as to the actual origins of Hizbullah. While many date its beginning in 1982, Marius Deeb places it earlier. See Marius Deeb, *Militant*

Islamic Movements in Lebanon: Origins, Social Basis, and Ideology, occasional papers (Washington, D.C.: Center for Contemporary Arab Studies, Georgetown University, 1986), pp. 13–14.

36. Shimon Shapira, "The Origins of Hizbullah," *Jerusalem Quarterly* 46, 1988: p. 130.

37. Augustus Richard Norton, "Lebanon: The Internal Conflict and the Iranian Connection," in John L. Esposito, ed., *The Iranian Revolution: Its Global Impact* (Miami: Florida International University Press, 1990), p. 23.

38. In contrast to many scholars, Marius Deeb regards Fadlallah as the political and religious leader of Hizbullah. See Marius Deeb, "Shia Movements in Lebanon," *Third World Quarterly,* p. 693.

39. Martin Kramer, "Muhammad Husayn Fadlallah," *Orient* 26:2 (1985): 147–48.

40. *The Middle East,* July 1989, p. 13.

41. Ibid.

42. Ibid.

43. "Iran Clamps Down on Hizbullah," *Christian Science Monitor,* October, 10, 1989, p. 3.

44. James P. Piscatori, "The Shia of Lebanon and Hizbullah: The Party of God," in Christie Jennett and Randal G. Stewart, eds., *Politics of the Future: The Role of Social Movements* (Melbourne: Macmillan, 1989), p. 301.

45. Fred Halliday, "Tunisia's Uncertain Future," *Middle East Report,* March–April 1990, p. 25.

46. "Nobody's Man—but a Man of Islam," in *The Movement of Islamic Tendency: The Facts* (Washington, D.C.: privately printed, 1987), p. 80. Similar comments appear in Abdelwahhab El-Effendi, "The Long March Forward," *Inquiry,* October 1987, p. 50.

47. "Nobody's Man," p. 84.

48. Interview with Ali Laridh, Tunis, 1989.

49. El-Effendi, "Long March," p. 51.

50. Interview with Ali Laridh, Tunis, 1989.

51. As quoted in Linda G. Jones, "Portrait of Rashid al-Ghannoushi," *Middle East Report,* July–August 1988, p. 20.

52. Clement Henry Moore, *Politics in North Africa* (Boston: Little, Brown, 1970), p. 108.

53. Dirk Vanderwalle, "From New State to the New Era: Toward a Second Republic in Tunisia," *Middle East Journal,* Autumn 1988, p. 612.

54. Dirk Vanderwalle, "Ben Ali's New Tunisia," *Field Staff Reports: Africa/Middle East,* 1989–90, no. 8.

55. Linda Jones, "Portrait," p. 22.

56. Vanderwalle, "From New State to the New Era," p. 603.

57. "Ben Ali Discusses Opposition Parties, Democracy," *FBIS-NES,* no. 29 (December 1989).

58. Halliday, "Tunisia's Uncertain Future," p. 26.

59. "The Autocrat Computes," *The Economist,* May 18, 1991, pp. 47–48.

60. *The Middle East,* September 1991, p. 18.

61. "Useful Plot," *The Economist,* June 1, 1991, p. 38.

62. "Amid Praise for Algerian System, Hopes for an Islamic Government," *The Message International* (August 1991), p. 16.

Chapter 6

1. When I first wrote this phrase, I feared that some might think it a bit outrageous. I was pleased to discover subsequently that the *National Review* published an article by Daniel Pipes entitled "The Muslims Are Coming! The Muslims Are Coming!" (November 19, 1990, pp. 28–31).

2. *Boston Globe,* July 27, 1991.

3. Patrick Bannerman, *Islam in Perspective: A Guide to Islamic Society, Politics and Law* (London: Routledge, 1988), p. 219.

4. "People Direct Islam in Any Direction They Wish," *Middle East Times,* May 28–June 3, 1991, p. 15.

5. Fouad Ajami, "Bush's Middle East Memo," *U.S. News and World Report,* December 26, 1988/January 2, 1989, p. 75.

6. "Can Islam, Democracy, and Modernization Co-Exist?," *Africa Report,* September–October 1990, p. 9.

7. I am indebted to Prof. R. K. Ramazani of the University of Virginia for this phrase.

8. "Can Islam, Democracy, and Modernization Co-Exist?," p. 9.

9. "Tunisia Warns of Islamic Radicals," *Washington Times,* October 25, 1991.

10. Ibid.

11. Bernard Lewis, "Roots of Muslim Rage," *Atlantic Monthly* 226:3 (September 1990): 47, 51.

12. Patrick J. Buchanan, "Rising Islam May Overwhelm the West," *New Hampshire Sunday News,* August 20, 1989.

13. Charles Krauthammer, "The New Crescent of Crisis: Global Intifada," *Washington Post,* February 16, 1990.

14. Buchanan, "Rising Islam."

15. Tariq Azim-Khan, a member of the Muslim Forum, as quoted in the *Independent,* October 29, 1991.

16. Dr. Hesham El-Essawy, chairman of the Islamic Society for the Promotion of Religious Tolerance, as quoted in the *Independent,* October 29, 1991.

17. Dominique Moisi of France's Institute of International Relations, as quoted in Judith Miller, "Strangers at the Gate: Europe's Immigration Crisis," *New York Times Magazine,* September 15, 1991, p. 86.

18. Ibid., p. 80.

19. Ibid., p. 86.

20. I am indebted to James Piscatori for reminding me about these important points.

21. *Atlantic Monthly,* September 1990, p. 2.

22. Lewis, "Roots," p. 56.

23. Ibid., p. 60.

24. James P. Piscatori, *Islam in a World of Nation States* (Cambridge: Cambridge University Press, 1986), p. 38.

25. Ibid.

26. Maxime Rodinson, *Europe and the Mystique of Islam* (Seattle: University of Washington Press, 1987), p. 67.

27. Lothrop Stoddard, *The New World of Islam* (New York: Scribner's, 1921), pp. 23–24.

28. As quoted in the *New York Times,* June 1, 1919, p. 9.

29. Raymond Aron, 'L'Incendie,' *L'Express,* December 1, 1979, as quoted in Piscatori, *Islam in a World of Nation States,* p. 39.

30. Cyrus Vance, *Hard Choices: Critical Years in American Foreign Policy* (New York: Simon and Schuster, 1983), pp. 408, 410, as quoted ibid.

31. Krauthammer, "The New Crescent."

32. Ibid.

33. "Exit Lebanon," *The New Republic,* November 12, 1990.

34. For a discussion of this point, see James P. Piscatori, *Islam in a World of Nation States,* p. 149.

35. "Democratization in the Middle East," *American–Arab Affairs* 36 (Spring 1991): 1–47; Michael C. Hudson, "After the Gulf War: Prospects for Democratization in the Arab World," *Middle East Journal* 45 (Summer 1991): 407–26.

36. For an analysis of this issue, see John L. Esposito and James P. Piscatori, "Democratization and Islam," *Middle East Journal* 45 (Summer 1991): 427–40.

37. Leslie Gelb, "The Free Elections Trap," *New York Times,* May 29, 1991.

38. Krauthammer, "The New Crescent," p. 2.

39. Buchanan, "Rising Islam."

40. *Middle East Times,* June 19–25, 1990.

41. Ibid.

42. John O. Voll, "For Scholars of Islam, Interpretation Need Not Be Advocacy," *The Chronicle of Higher Education,* March 22, 1989, p. A48. See also Daniel Pipes, *The Rushdie Affair: The Novel, the Ayatollah, and the West* (New York: Birch Lane Press, 1990).

43. For a sampling of international reactions, see Lisa Appignanesi and Sara Maitland, eds., *The Rushdie File* (Syracuse, N.Y.: Syracuse University Press, 1990). Parenthetical references in the text are to Salman Rushdie, *The Satanic Verses* (New York: Viking, 1989).

44. James P. Piscatori, "The Rushdie Affair and the Politics of Ambiguity," *International Affairs* 66 (1990): 767–89.

45. "Repeating an Apostasy Is an Apostasy: Religious Leaders Urge Muslims to Ignore Rushdie Issue," *Middle East Times,* March 7–13, 1989, p. 14.

46. *Law, Blasphemy and the Multi-Faith Society: Report of a Seminar,* Discussion Papers 1 (London: Commission of Racial Equality, February 1990), p. 53.

47. See John L. Esposito, "Jihad in a World of Shattered Dreams: Islam, Arab Politics, and the Gulf Crisis," *The World & I,* February 1991, pp. 515–27.

48. James P. Piscatori, ed., *Islamic Fundamentalisms and the Gulf Crisis* (Chicago: American Academy of Arts and Sciences, 1991); and Yvonne Y. Haddad, "Operation Desert Shield/Desert Storm: The Islamist Perspective," in Phyllis

Bennis and Michel Moushabeck, eds., *Beyond the Storm: The Gulf Crisis Reader* (New York: Olive Branch Press, 1991), ch. 23.

49. North Africa (Algeria, Morocco, Tunisia, Mauretania, and the Sudan) witnessed large demonstrations. *Washington Post,* September 14, 1990, and *Los Angeles Times,* September 22, 1990.

50. "Islam Divided," *The Economist,* September 22, 1990, p. 47.

51. Abdurrahman Alamoudi in *The Washington Report on Middle East Affairs,* October 1990, p. 69.

52. *Herald,* September 1990, p. 30.

53. "The Gulf Crisis," *Aliran* 10:8 (1990): 32.

54. "Exposing U.S. Motives: A Third World View," *Aliran* 10:8 (1990): 38.

55. This is true even in the eyes of many Saudis and Egyptians. See *New York Times,* October 24, 1990, and *Middle East Times,* October 16–22, 1990.

56. *Dawn,* February 1, 1991.

57. *Christian Science Monitor,* January 24, 1991.

58. *Far Eastern Economic Review,* January 24, 1991.

59. *The Economist,* January 26, 1991.

60. See for example Daniel Lerner, *The Passing of Traditional Society: Modernizing the Middle East* (New York: The Free Press, 1958), and Manfred Halpren, *The Politics of Social Change in the Middle East and North Africa* (Princeton, N.J.: Princeton University Press, 1963). For an analysis and critique of the factors which influenced the development of modernization theory, see Fred R. von der Mehden, *Religion and Modernization in Southeast Asia* (Syracuse, N.Y.: Syracuse University Press, 1988).

61. See for example Harvey Cox, *The Secular City—Urbanization and Secularization in Theological Perspective* (New York: Macmillan, 1965), and *Religion in the Secular City: Toward a Postmodern Theology* (New York: Simon and Schuster, 1984); Dietrich Bonhoeffer, *Letters and Papers from Prison,* rev. ed. (New York: Macmillan, 1967); and William Hamilton and Thomas Altizer, *Radical Theology and the Death of God* (Indianapolis, Ind.: Bobbs-Merrill, 1966).

62. See for example the influential introduction of H. A. R. Gibb, *Mohammedanism* (Oxford: Oxford University Press, 1962), or John B. Christopher, *The Islamic Tradition* (New York: Harper and Row, 1972).

63. Edward W. Said, *Orientalism* (New York: Vintage Books, 1979). For coverage of Islam specifically by the media and experts, see his *Covering Islam* (New York: Pantheon, 1981).

64. Documentation and reports of human rights violations have been numerous. See the U.S. State Department's *Country Reports on Human Rights Practices* for 1991 (Washington, D.C.: U.S. Government Printing Office, 1992), pp. 1334ff and 1615ff; "Tunisia: Thousands Held Illegally, Torture Routine in Crackdown on Islamic Opposition" (Washington, D.C.: Amnesty International, March 3, 1992); and *Tunisia: Prolonged Incommunicado Detention and Torture* (Washington, D.C.: Amnesty International, March 1992). Sudan's Islamically oriented government has received similar criticism. See, for example, "Sudan: The Ghosts Remain" (New York: Africa Watch Report, April 1992) and *Sudan: A Human Rights Disaster* (New York: Africa Watch, March 1990).

SELECT BIBLIOGRAPHY

For a comprehensive bibliography on the role of Islam in contemporary history and politics, see Yvonne Yazbeck Haddad, John Obert Voll, and John L. Esposito, eds., *The Contemporary Islamic Revival: A Critical Survey and Bibliography* (New York: Greenwood Press, 1991).

Abrahamian, Ervand. *Iran Between Two Revolutions.* Princeton, N.J.: Princeton University Press, 1982.

Ahmed, Akbar S. *Postmodernism and Islam: Predicament and Promise.* London: Routledge, 1992.

Ajami, Fouad. *The Arab Predicament.* New York: Cambridge University Press, 1982.

———. *The Vanished Imam: Musa Sadr and the Shia of Lebanon.* Ithaca, N.Y.: Cornell University Press, 1986.

Akhavi, Sharough. *Religion and Politics in Contemporary Iran.* Albany: SUNY Press, 1980.

Al-e-Ahmad, Jalal. *Gharbzadegi.* Lexington, Ky.: Mazda, 1982.

Algar, Hamid, trans. *Islam and Revolution: Writings and Declarations of Imam Khomeini.* Berkeley, Calif.: Mizan Press, 1981.

Antoun, Richard T., and Mary Elaine Hegland, eds. *Religious Resurgence: Contemporary Cases in Islam, Christianity, and Judaism.* Syracuse, N.Y.: Syracuse University Press, 1987.

Arjomand, Said Amir. *The Turban for the Crown: The Islamic Revolution in Iran.* New York: Oxford University Press, 1988.

———, ed. *Authority and Political Culture in Islam.* Albany: SUNY Press, 1988.

Ayoob, Mohammed, ed. *The Politics of Islamic Reassertion.* New York: St. Martin's Press, 1981.

Ayoub, Mahmoud M. *Islam and the Third Universal Theory: The Religious Thought of Muammar al-Qadhdhafi.* London: KPI, 1987.

Ayubi, Nazih. *Political Islam: Religion and Politics in the Arab World.* London: Routledge, 1991.

Baker, Raymond. *Sadat and After: Struggles for Egypt's Political Soul.* Cambridge, Mass.: Harvard University Press, 1990.

Bakhash, Shaul. *The Reign of the Ayatollahs: Iran and the Islamic Revolution.* New York: Basic Books, 1984.

Bannerman, Patrick. *Islam in Perspective.* London: Routledge, 1988.

Banuazizi, Ali, and Myron Weiner, eds. *The State, Religion, and Ethnic Politics.* Syracuse, N.Y.: Syracuse University Press, 1986.

Bennigsen, Alexandre, and Marie Boxup. *The Islamic Threat to the Soviet State.* New York: St. Martin's Press, 1983.

Bill, James A. *The Eagle and the Lion.* New Haven, Conn.: Yale University Press, 1988.

Binder, Leonard. *Islamic Liberalism: A Critique of Development Ideologies.* Chicago: University of Chicago Press, 1988.

Burke, Edmund, and Ira M. Lapidus, eds. *Islam, Politics, and Social Movements.* Berkeley: University of California Press, 1988.

Cole, Juan R.I., and Nikki R. Keddie, eds. *Shiism and Social Protest.* New Haven, Conn.: Yale University Press, 1986.

Curtis, Michael, ed. *Religion and Politics in the Middle East.* Boulder, Colo.: Westview Press, 1982.

Daniel, Norman. *Islam: Europe and Empire.* Edinburgh: Edinburgh University Press, 1966.

Dekmejian, R. Hrair. *Islam in Revolution: Fundamentalism in the Arab World.* Syracuse, N.Y.: Syracuse University Press, 1985.

Dessouki, Ali E. Hillal, ed. *Islamic Resurgence in the Arab World.* New York: Praeger, 1982.

Djait, Hichem. *Europe and Islam: Cultures and Modernity.* Berkeley: University of California Press, 1985.

Donohue, John J., and John L. Esposito, eds. *Islam in Transition: Muslim Perspectives.* New York: Oxford University Press, 1982.

Eickelman, Dale F. *The Middle East: An Anthropological Approach.* 2nd ed. Englewood Cliffs, N.J.: Prentice Hall, 1989.

Enayat, Hamid. *Modern Islamic Political Thought.* Austin: University of Texas Press, 1982.

Esposito, John L. *Islam and Politics.* 3rd ed. Syracuse, N.Y.: Syracuse University Press, 1991.

———. *Islam in Asia: Religion, Politics and Society.* New York: Oxford University Press, 1987.

———. *Islam: The Straight Path.* Expanded ed. New York: Oxford University Press, 1991.

———, ed. *The Iranian Revolution: Its Global Impact.* Miami: Florida International University Press, 1990.

———, ed. *Voices of Resurgent Islam.* New York: Oxford University Press, 1983.

Fischer, Michael M.J. *Iran: From Religious Discourse to Revolution.* Cambridge, Mass.: Harvard University Press, 1980.

Gibb, Hamilton A.R., and Harold Bowen. *Islamic Society and the West.* Oxford: Oxford University Press, 1960.

Gilsenan, Michael. *Recognizing Islam: Religion and Society in the Modern Arab World.* New York: Pantheon, 1983.

Goldschmidt, Arthur, Jr. *A Concise History of the Middle East.* 3rd ed. Boulder, Colo.: Westview Press, 1991.

Haddad, Yvonne Y. *Contemporary Islam and the Challenge of History.* Albany: SUNY Press, 1982.

———,ed. *The Islamic Impact.* Syracuse, N.Y.: Syracuse University Press, 1984.

Hiro, Delip. *Holy Wars: The Rise of Islamic Fundamentalism.* London: Routledge, 1989.

Hodgson, Marshall G.S. *The Venture of Islam.* 3 vols. Chicago: University of Chicago Press, 1974.

Hourani, Albert. *Arabic Thought in the Liberal Age, 1798–1939.* New York: Cambridge University Press, 1983.

———. *Europe and the Middle East.* Berkeley: University of California Press, 1980.

———. *A History of the Arab Peoples.* Cambridge, Mass.: Harvard University Press, 1991.

Hudson, Michael C. *Arab Politics: The Search for Legitimacy.* New Haven, Conn.: Yale University Press, 1977.

Hunter, Shireen T., ed. *The Politics of Islamic Revivalism: Diversity and Unity.* Bloomington: Indiana University Press, 1988.

Islamic Fundamentalism and Islamic Radicalism. Committee on Foreign Affairs, House of Representatives, Washington, D.C.: U.S. Government Printing Office, 1985.

Johnson, Nels. *Islam and the Politics of Meaning in Palestinian Nationalism.* London: Kegan Paul, 1983.

Keddie, Nikki R. *Iran: Religion, Politics and Society.* London: Frank Cass, 1980.

———. *Roots of Revolution: An Interpretive History of Modern Iran.* New Haven, Conn.: Yale University Press, 1981.

Kepel, Gilles, *Muslim Extremism in Egypt: The Prophet and Pharaoh.* Berkeley: University of California Press, 1986.

Kramer, Martin. *Political Islam.* Beverly Hills, Calif.: Sage Publications, 1980.

———. *Shiism, Resistance and Revolution.* Boulder, Colo.: Westview Press, 1987.

Lapidus, Ira M. *A History of Islamic Societies.* Cambridge: Cambridge University Press, 1988.

Lawrence, Bruce. *Defenders of God: The Fundamentalist Revolt Against the Modern Age.* New York: Harper and Row, 1989.

Lewis, Bernard. *The Middle East and the West.* New York: Harper and Row, 1968.

———. *The Muslim Discovery of Europe.* New York: W. W. Norton, 1982.

———. *The Political Language of Islam.* Chicago: University of Chicago Press, 1988.

Mayer, Ann Elizabeth. *Islam and Human Rights: Tradition and Politics.* Boulder, Colo.: Westview Press, 1991.

Mitchell, Richard. *The Society of Muslim Brothers.* New York: Oxford University Press, 1969.

Mortimer, Edward. *Faith and Power.* New York: Random House, 1982.

Mottahedeh, Roy P. *The Mantle of the Prophet: Religion and Politics in Iran.* New York: Random House, Pantheon, 1986.

Munson, Henry. *Islam and Revolution in the Middle East.* New Haven, Conn.: Yale University Press, 1988.

Norton, Augustus Richard. *AMAL and the Shia: Struggle for the Soul of Lebanon.* Austin: University of Texas Press, 1987.

Peters, Rudolph. *Islam and Colonialism: The Doctrine of Jihad in Modern History.* The Hague: Mouton, 1979.

Pipes, Daniel. *The Rushdie Affair: The Novel, the Ayatollah, and the West.* New York: Birch Lane Press, 1990.

Piscatori, James P. *Islam in a World of Nation States.* Cambridge: Cambridge University Press, 1986.

———, ed. *Islam in the Political Process.* Cambridge: Cambridge University Press, 1983.

———, ed. *Islamic Fundamentalisms and the Gulf Crisis.* Chicago: University of Chicago Press, 1991.

Rahman, Fazlur. *Islam and Modernity.* Chicago: University of Chicago Press, 1982.

Rodinson, Maxime. *Europe and the Mystique of Islam.* Seattle: University of Washington Press, 1987.

Roff, William, ed. *Islam and the Political Economy of Meaning.* London: Croom Helm, 1987.

Roy, Olivier. *Islam and Resistance in Afghanistan.* Cambridge: Cambridge University Press, 1986.

Sahliyeh, Emile, ed. *Religious Resurgence and Politics in the Modern World.* Albany: SUNY Press, 1990.

Said, Edward W. *Covering Islam.* New York: Pantheon, 1981.

———. *Orientalism.* New York: Vintage, 1979.

Shariati, Ali. *On the Sociology of Islam.* Berkeley, Calif.: Mizan Press, 1979.

Sivan, Emmanuel. *Radical Islam: Medieval Theology and Modern Politics.* New Haven, Conn.: Yale University Press, 1985.

Sivan, Emmanuel, and Menachem Friedman, eds. *Religious Radicalism and Politics in the Middle East.* Albany: SUNY Press, 1990.

Smith, Donald E., ed. *Religion and Political Modernization.* New Haven, Conn.: Yale University Press, 1974.

Sonn, Tamara. *Between Quran and Crown.* Boulder, Colo.: Westview Press, 1990.

Southern, R.W. *Western Views of Islam in the Middle Ages.* Cambridge, Mass.: Harvard University Press, 1962.

Stowasser, Barbara, ed. *The Islamic Impulse.* London: Croom Helm, 1987.

Voll, John Obert. *Islam, Continuity and Change in the Modern World.* Boulder, Colo.: Westview Press, 1982.

Watt, W. Montgomery. *Islamic Political Thought.* New York: Columbia University Press, 1980.

Weiss, Anita, ed. *Islamic Reassertion in Pakistan: The Application of Islamic Laws in a Modern State.* Syracuse, N.Y.: Syracuse University Press, 1986.

Wright, Robin. *Sacred Rage: The Wrath of Militant Islam.* New York: Simon and Schuster, 1985.

———. *In the Name of God: The Khomeini Decade.* New York: Simon and Schuster, 1989.

Index

Note: Arabic names beginning with "al" or other particles are alphabetized according to the element following this particle. For example, al-Farag, Muhammad, is found under Fatag, with the particle "al" retained in front of it.

Abbasid Caliphate, 31, 40, 43, 49
Abduh, Muhammad, 60, 61, 64; and modernism, 201, 55, 57–58, 165
Abdulhamid II, 54, 56
ABIM (Malaysian Youth Movement), 20, 187
al-Afghani, Jamal al-Din, 60, 61, 62, 64, 65; and modernism, 55, 56–57, 58
Afghanistan, 11, 70, 183
Africa, 51, 86, 87, 176 (see also North Africa; South Africa; individual countries); Islam in, 4, 37, 49, 70, 175
Afwaj al-Muqawimah al-Lubnaniyah. See AMAL
Ahmad, Qazi Hussein, 196
Ahmad Khan, Sir Sayyid, 55, 58–59, 60, 61
Alexius I, 40, 41
Algeria, 55, 68, 70, 154 (see also Maghreb; North Africa); government in, 152, 154, 184; and the Gulf War, 170, 195, 196, 197; Islamic modernism in, 55; Islamic revivalism in, 11, 152, 154, 162, 165, 166, 169, 171, 198, 206, 208; nationalism in, 62, 65, 70; political reform in, 162, 165, 166, 167, 184, 185, 187, 189, 206, 208, 211; revolution in, 80, 107
Algiers, 52, 181
Ali (Muhammad's cousin), 30, 145
Allah, Hajji Shariat, 50
Alliance of Nonaligned Nations, 72
Ally, Mashuq, 192
AMAL (Lebanese Resistance Battalions),

14, 143, 144–45; compared to Hizbullah, 22, 148, 149–51; conflict with Hizbullah of, 150–51; goals of, 22, 145, 149–50; and Iran, 151; and Israel, 150–51; schism in, 146, 150; and Syria, 151; and the U.S., 150–51; and the West, 150, 151, 166
Amin, Idi, 86
Amin, Qasim, 58
Anatolia, 38, 44
Ansar, 88
Anti-imperialism, 66, 74, 81, 121 (see also European colonization: criticism of); and Islamic modernism, 56, 57, 58, 59, 62; and Nasser, 71, 72; and pan-Islam, 181–82; and Sadat, 97; and Saddam Hussein, 194
Arab Christian Literary Movement, 63
Arab Human Rights Organization, 99
Arab League, 117, 193
Arab nationalism, 63–64 (see also Nasserism; individual states); failure of, 15, 47, 75, 76, 79, 94, 154, 183; in The Green Book; 81; and the Gulf War, 194, 196, 197; and Islam, 69, 74–76; and Israel, 73–74, 75; in Lebanon, 145; and 1967 Arab-Israeli war, 17, 154; program of, 72–73, 183; and secularism, 15, 69, 74, 75, 78–79; and the West, 69, 70–73
Arabia, 36, 182 (see also Saudi Arabia); Christianity in, 28, 31, 33, 39; and Muhammad, 27, 28, 30, 31
Arab-Israel War (1948), 72, 73, 120, 170

Arab-Israeli war (1967), 133, 170; and
  Arab nationalism, 17, 75, 76, 79, 154;
  and Arab-Muslim identity, 12, 13, 14,
  76, 94, 154
Army of God (Jund Allah); of Egypt, 95;
  and Lebanon, 148; and terrorism, 4,
  119, 206
Aron, Raymond, 182
Asia: and democracy, 211; Islam in, 12,
  13, 175, 176; South (*see* South Asia);
  Southeast (*see* Southeast Asia)
Association of Algerian Ulama, 65
Aswan Dam, 72
Ataturk, Kemal, 66, 78, 153
Atlantic Monthly, 173, 174, 177
Averroës, 34
Avicenna, 34
Azerbaijan, 181, 182, 183

Baath party, 72, 75, 78–79; in Egypt, 72,
  73; in Iraq, 20, 70, 73; in Syria, 70
Badis, Abd al-Hamid Ibn. *See* Badis, Ben
Badis, Ben, 55, 65, 152
Bahrain, 22, 114, 184
Balkans, 43, 44, 63, 183
Bangladesh, 12, 13, 15, 67, 70; and the
  Gulf War, 196, 197
Bani-Sadr, 113
al-Banna, Hassan, 108, 128, 137;
  assassination of, 70, 129, 130;
  ideological worldview of, 122–23, 126;
  influence of, 70, 130, 134, 152, 154, 155;
  and modernism, 121, 124–25; and
  modernization, 121, 122, 125; Muslim
  Brotherhood founded by, 120; West
  criticized by, 121–22, 124, 125, 129
Bannerman, Patrick, 169
Baqir al-Sadr, Ayatollah Muhammad, 20
al-Bashir, Omar, 92–93, 117, 204
Bazargan, Mehdi, 105–6, 110, 113
Bedouin, 28, 31, 32
Ben Ali, Zeine Abedin, 161–63
Benhadj, Ali, 167, 189
Bennabi, Malek, 152, 155
Berri, Nabih, 145, 146, 150
Bhutto, Benazir, 192
Bhutto, Zulfikar Ali, 14, 79, 162
Bill, James, 104
Black Friday, 111
Bodin, Jean, 43
Bokassa, Jean-Bédel, 86
Bosworth, C. E., 42
"Boundary Between Religion and Solcial
  Affairs" (Bazargan), 105–6
Bourguiba, Habib, 79, 155, 156, 162; and

the Islamic Tendency Movement, 158–
  59, 160–62; and the modernization of
  Tunisia, 68, 152, 153
Brelwi, Ahmad, 50
Britain, 69 (*See also similar subheadings
  under* West); and Egypt, 19, 48, 52, 56,
  71, 72, 120; and the Gulf, 51, 104; India
  colonized by, 51, 53, 58–59, 60; and
  Iran, 19, 57, 65, 102, 103, 104, 171; Iraq
  colonized by, 51, 63; Israel supported
  by, 19, 120; and Jordan, 63; and
  Lebanon, 19, 141; and Libya, 80; and
  Malaysia, 51, 68; mandate countries
  created by, 75, 121; Muslim Parliament
  of, 176; Muslims in, 176; and Pakistan,
  68; Palestine colonized by, 51, 63; and
  Rushdie affair, 190, 192; and South
  Asia, 66, 68; and the Sudan, 90;
  Transjordan colonized by, 51
Buchanan, Patrick, 175, 177, 188
Buddhism, 37, 198, 199
Bugeaud, Marshal, 52
Bush, George, 72, 91, 195
Byzantine empire, 3, 31, 38, 39, 40, 42

Caetani, Leone, 181
Caliphate Movement, 66, 120
Camp David Accords, 88, 94, 136;
  opposition to, 95, 97, 99, 117, 132, 139
Capitalism: rejected by Muslims, 15, 59,
  69, 71, 72, 73, 81, 125, 179; versus
  communism, 4, 70, 172
Carter, Jimmy, 93
Central African Republic, 86
Central Intelligence Agency, 104
China, 72, 81; Islam in, 4, 11, 32, 181
Chinese-Malay riots, 12, 13
Christianity, 192, 199 (*see also individual
  states*); in Arabia, 28, 31, 33, 39;
  contrasted with Islam, 34, 127, 174, 208;
  in Egypt, 97, 98, 132, 188; historic
  conflict with Muslims, 3–4, 24, 44, 46,
  51, 52, 170, 172, 175, 181, 188, 205 (*see
  also* Crusades; Ottoman empire); and
  the Intifada, 182; Islam as threat to, 5,
  25, 32, 37–38, 45–46, 47–48, 180; in
  Islamic states, 14, 16, 80, 88, 97, 98,
  132, 182, 188; and Jerusalem, 40–41;
  and Judaism, 38, 81; missionaries of, 48,
  49, 52–53; and modernization, 48;
  origins of, and Islam, 25, 26, 30, 61; and
  Ottoman empire, 42, 43–44, 179;
  similarities to Islam, 10, 25, 33, 37–38,
  44, 59, 81, 172, 188–89; and Sufism, 37;

and terrorism, 174, 180, 198; as threat
to Islam, 49, 52–53, 69, 92, 135, 177
CIA. *See* Central Intelligence Agency
Cold War, 4, 5, 151, 177; and Arab
nationalism, 70, 71; and Islamic
revivalism, 75–76
Colonialism. *See* European colonization
Communism, 5, 73, 87, 198, 210 (*See also*
Marxism); collapse of, 4, 182, 183 (*see
also* Soviet Union, breakup of); and
Islamic revivalism, 85, 125, 127, 134,
168, 208; versus capitalism, 4, 70, 172
Constantinople, 40, 42, 43
Constitution party. *See* Destour party
Constitutional Democratic Rally (Tunisia),
162
Constitutional Revolution (Iran, 1905–
11), 65, 102, 110
Crecelius, Daniel, 9–10
Cromer, Lord, 48, 51
Crusades, 3, 39–42, 199; influencing
Muslim views of West, 19, 40, 42, 46,
51, 52, 92, 134, 135, 171, 179, 190;
Western myths concerning, 39–40, 42;
and Western view of Islam, 42, 43, 44,
45, 170, 175, 180
Cyrus the Great, 104

al-Dawa (Islamic Call Society, Iraq), 20
Deeb, Marius, 222n.35, 222n.38
Democracy: in Asia, 211; and Christianity,
170; in Eastern Europe, 4, 169–70; and
Islam, 3, 6, 185–89, 207, 212; and Islamic
activists, 167, 179, 184–87, 188, 189, 210,
211, 212; in Middle East, 76, 169–70,
184–85, 187, 189, 208, 211; in Soviet
Union, 169–70, 184, 211; and terrorism,
198, 205; in Tunisia, 187, 211; Western,
68, 126, 170, 186, 187, 190, 207, 211
Destour party (Tunisia), 65, 156
al-Dhahabi, Husayn, 134
al-Din, Nasir, 56

Eastern Europe, 32, 85, 208;
democratization movement in, 4, 169–
70, 183, 184, 211
Eden, Anthony, 72
Egypt, 15, 97; and Britain, 19, 48, 52, 56,
71, 72, 120; and China, 72; and early
Islamic expansionism, 38; and France,
19, 56, 72; government in, 184, 185,
187; and the Gulf War, 194, 195, 196,
197; and India, 72; and Indonesia, 72;
and Iran, 94, 95; and Iranian
Revolution, 115; Islam in, 12, 20, 23,
79, 96, 98–99, 117, 133, 135; Islamic
modernism in, 55, 56, 57, 59, 60–61, 64;
Islamic revivalism in, 10, 11, 64, 93–
100, 126, 132–40, 151, 154, 164, 165,
166, 169, 185, 187, 198, 203, 206; and
Israel, 13, 17, 72, 73, 94, 98, 120, 139,
170 (*see also* Arab-Israeli war [1967]);
Camp David Accords); and Libya, 79,
85, 86, 117; modernization and
secularism in, 9–10, 23, 54, 61, 78–79,
94, 164; nationalism in, 56, 61, 62, 63–
64, 70, 75, 78–79, 94, 182; revolution in,
79, 80; and Rushdie affair, 191; and
Soviet Union, 72, 97; Sufis of, 94, 100,
132, 139; and the Sudan, 79, 87, 91; and
Syria, 73; terrorism in, 133; ulama in,
100; and U.S., 19, 72, 95, 98, 117, 139,
184; and the West, 19, 68–69, 95, 97, 98,
99, 139
Egyptian-Israeli war (1973), 17, 94
Eisenhower, Dwight, 146
Emancipation of Women, The (Amin), 58
Emirates. *See* United Arab Emirates
Ennahda. *See* Islamic Tendency
Movement; Renaissance Party
Erasmus, 44
Ershad, Muhammad, 12
Europe, 41, 75, 87, 191, 196 (*see also
similar subheadings under* West); and
boundaries of Arab world, 72, 75, 121;
colonial expansion of (*see* European
colonization; West: imperialism of); and
early Islamic expansionism, 34, 38, 46;
Eastern (*see* Eastern Europe); and Iran,
101, 102, 107, 191; Islam viewed in, 44–
46, 47–48; Muslims in, 4, 13, 175–77;
and Ottoman empire, 3, 42–44, 170;
and the U.S., 169, 196, 208
European colonization, 3, 47–49, 51–53,
180 (*see also* West: imperialism of);
criticism of, 59, 71, 85, 120, 121, 186
(*see also* Anti-imperialism); and
democracy in Middle East, 185; end of,
70, 75, 183; and Israel, 73; and
modernization, 8–9, 15, 51; and Muslim
identity, 17, 46, 49, 51, 52; Muslim
responses to, 51, 53–62, 66, 69, 170,
171–72; and Nasser, 72; and Pan-Islam,
181; and Western-Islamic relations
today, 19, 42, 46, 47
Excommunication and Emigration. *See*
Takfir wal-Hijra

Fadlallah, Muhammad Husayn, 148–49,
150, 222n.38

Fahd, King of Saudi Arabia, 8, 117
Faisal, King of Saudi Arabia, 85
Fanaticism, or Muhammad the Prophet
   (Voltaire), 46
al-Farabi, 34
al-Farag, Muhammad, 96–97, 134, 137
Faraidiyyah, 50
Farouk, King of Egypt, 71, 129
al-Fasi, Allal, 55, 64
FIS. *See* Islamic Salvation Front
Four Rightly Guided Caliphs, 31
France, 41, 75, 108, 121, 182 (*see also
   similar subheadings under* West); Africa
   colonized by, 51, 52, 64, 65, 68; colonial
   expansion of, 48, 51, 197 (*see also*
   European colonization); and Egypt, 19,
   56, 72; and Iran, 19, 51, 104; and Iraq,
   63; Israel supported by, 19; and Jordan,
   63; and Lebanon, 19, 51, 63, 68, 141,
   148, 150; Muslims in, 171, 176–77; and
   Palestine, 63; Syria colonized by, 51, 63;
   and Tunisia, 68, 153, 154, 155, 158, 161
Free Officers: Egypt, 71, 74; Sudan, 87
Fundamentalism, 3, 4 (*see also* Islamic
   revivalism and activism); definitions of,
   7–8
Future of Culture in Egypt (Husayn), 60–
   61

Gambia, 86
Garang, John, 90
Gaza Strip: and the Gulf War, 196; Islam
   in, 4, 11; Israeli capture and occupation
   of, 12, 13, 171, 182–83, 195–96; and the
   U.S., 207 (*see also* Israel: and the U.S.)
GCC. *See* Gulf Cooperation Council
Gemayel, Amin, 146
General Union of Tunisian Workers
   (UGTT), 156, 158
Germany, 63, 66
Ghannoushi, Rashid, 166; and Islamic
   revivalism, 154, 155–58, 159, 160–61,
   167; West criticized by, 156, 165
Gharbzadegi (Jalal-e-Ahmad), 105–6
al-Ghazzali, Muhammad, 155
Gobtzadeh, Sadeq, 113
Golan Heights, 12, 196
Gorbachev, Mikhail S., 191
Gore, Albert, 183
Green Book, The (Qaddafi), 81–82, 83, 84
Green Menace, 5
Gregevich, Bartholomew, 43
Gulf Cooperation Council (GCC), 21
Gulf crisis of 1990–91. *See* Gulf War
Gulf states (*see also individual states*); and

Britain, 51, 104; and Iran, 22, 183; and
   Iraq, 21, 183; Islamic revivalism in, 70,
   132, 203; Shii in, 20, 22, 114; Sunni in,
   20, 21; and the U.S., 21, 114
Gulf War (1990–91), 5, 23–24, 46, 163,
   167, 184, 185, 193–97

al-Hakim, Muhsin, 147
Hamas, 182–83
ul-Haq, Zia, 130, 204, and Islam, 8, 12,
   79, 117, 163
Harb, Raghib, 148
Hassan, King of Morocco, 152
Higgins, Col. William R., 150–51
High Caliphal Period, 34
Hijra, 29, 53
Hinduism, 37, 180, 198, 199; in India, 66,
   67, 121
Hizb al-Nahda. *See* Renaissance Party
Hizbullah, 84; compared to AMAL, 148,
   149–51; conflict with AMAL of, 150–
   51; and France, 148, 150, 161; goals of,
   14, 148, 149–50; groups encompassed
   by, 148; and Iran, 14, 22, 145, 146, 147,
   148, 149–50, 151; and Israel, 145, 146,
   150–51; and Kuwait, 148; leadership of,
   147, 148–49, 222n.38; organization of,
   148–49; rise of, 145–48, 222n.35; and
   Saudi Arabia, 148; and Syria, 151; and
   terrorism, 4, 11, 119, 150, 151, 206; and
   the U.S., 146, 148, 150–51; and the
   West, 150, 166
Holy War. *See* Jihad
Holy War Society. *See* Jamaat al-Jihad
Hourani, Albert, 45, 48
Husain, 110, 111, 145
Husain, Shii Imam, 143
Husayn, Taha, 59, 60–61, 64
Hussein, King of Jordan, 68
Hussein, Saddam, 142; call for jihad
   against foreigners by, 3, 193, 195; and
   Gulf War, 4, 193–94, 195, 196, 197;
   denounced, 4, 20, 21; and Iranian
   Revolution, 20; and Islam, 23–24, 193,
   194; and the West, 72, 170, 179, 194

Ibn-Hazm, 152
Ibn-Rushd, 34
Ibn-Sina, 34
Ibn-Taymiyya, 152
Ibrahim, Saad Eddin, 132–33
Idris, King of Libya, 80
Ikhwan al-Muslimin. *See* Muslim
   Brotherhood
ILO. *See* Islamic Liberation Organization

IMF. *See* International Monetary Fund
In the Shade of the Quran (Qutb), 128
Independence party. *See* Istiqlal party
India, 42, 72 (*See also* South Asia); British
  colonization of, 51, 53, 58–59, 60; and
  the Gulf War, 196; Hindus in, 66, 67,
  121; Islamic modernism in, 55, 56, 58–
  60; Islamic revivalism in, 11, 50, 70;
  Muslims in, 13, 184; nationalism in, 60,
  66–67; and Rushdie affair, 191
Indian Freedom (Independence)
  Movement, 121
Indian National Congress, 66
Indonesia, 72, 120; Dutch colonization of,
  51; and the Gulf War, 195, 196, 197; and
  Iranian Revolution, 115; Islamic
  modernism in, 61; and Islamic
  revivalism, 11, 50, 115, 130; and Libya,
  86; Muslims in, 13; nationalism in, 62;
  and the U.S., 195; and the West, 115,
  195
Inquisition, 3, 171, 180
International Monetary Fund (IMF), 88,
  90, 91
Intifada, 4, 182–83, 184, 196
Iqbal, Muhammad, 62, 66, 155; and
  Islamic modernism, 55, 59–60, 61, 62,
  165, 201
Iran, 11, 42, 56, 188; and Bahrain, 22; and
  Britain, 19, 51, 65, 102, 103, 104, 171;
  and Egypt, 94, 95; and France, 19, 51,
  104; government in, 204, 205; and Gulf
  states, 22, 183; and the Gulf War, 194,
  196, 197; hostages taken by, 170, 182,
  192; and Iraq, 22, 183 (*see also* Iran-Iraq
  war); Islam in, 8, 79, 89, 101–3, 109–11,
  115, 116, 140–41, 143, 182; Islamic
  revivalism in, 8, 10, 11, 23, 164, 187–88,
  203, 209; and Israel, 22, 109; and
  Lebanon, 22, 116, 141, 142, 146–47,
  148, 149–50, 151; modernization in, 11,
  18, 23, 54, 94, 101, 103, 105, 115, 164;
  nationalism in, 62, 63, 65–66, 102; non-
  Muslims in, 188, 189; and Rushdie
  affair, 190–93; and Saudi Arabia, 22,
  184; SAVAK of, 104, 106; Shii Islam in
  (*see* Iran: Islam in); and the Soviet
  Union, 22, 51, 65, 103, 108, 171, 191;
  and the Sudan, 92; and Syria, 147, 151;
  and terrorism, 5, 8, 22, 113, 114, 170,
  180; and Tunisia, 159, 160, 161; ulama
  in, 102, 106, 107–8, 109, 112, 113, 114;
  and the U.S., 11, 16, 18, 19, 22, 68, 101,
  104, 108, 116, 117, 171, 178, 182, 184,
  191–92, 197, 198, 204, 207; and the

West, 8, 16, 18, 19, 22, 101, 102, 103–4,
  105, 106, 107, 108–9, 111, 116, 140, 184,
  191–92, 204
Iranian Revolution (1978–79), 14, 46, 93,
  110, 171; export of, 19–22, 113–15, 116,
  151, 159, 170, 180, 209; impact on the
  Muslim world of, 18, 94, 113–15, 144,
  145, 147, 149, 155, 156, 157, 166, 182;
  and Islam, 4, 12, 18, 19–20, 145, 209;
  and Khomeini, 20, 108, 114, 170; and
  the West, 4, 17, 18, 113–14, 192, 198,
  207
Iran-Iraq war, 20, 21, 183, 203, 209;
  effects on Iran of, 115, 151, 191
Iraq, 38, 68, 108, 132, 171; Baathism in,
  20, 70, 73; British colonization of, 51,
  63; and France, 63; and Gulf states, 21,
  183; Hashemites of, 71; and Iran, 22,
  183 (*see also* Iran-Iraq war); and Iranian
  Revolution, 20, 115; Islam in, 183; and
  Israel, 196; Kuwait invaded by (*see* Gulf
  War); and Lebanon, 142, 147;
  nationalism in, 22, 70; and Palestine,
  197; revolution in, 80; Shii of, 20, 22;
  and Syria, 183; and the U.S. (*see* Gulf
  War); and the West, 115 (*see also* Gulf
  War)
Islam; branches of (*see* Shii Islam; Sunni
  Islam); in caliphate period, 31–32; and
  Christianity (*see* Christianity);
  community-state of (*see* Ummah);
  confrontational pattern of, 177–78; and
  democracy (*see* Democracy: and Islam);
  dependence on West of, 81, 186, 194,
  206; and diversity, 204–5, 207;
  expansionism of, 18, 31–32, 33–34, 37,
  38–39, 45, 50, 85, 170 (*see also*
  Crusades; Ottoman empire); in
  fundamentalism in (*see*
  Fundamentalism; Islamic revivalism and
  activism); in global politics, 4, 10, 11–
  12, 17; laws of (*see* Sharia); missionaries
  of, 37, 38, 86; mystical tradition of (*see*
  Sufism); and non-Muslims, 38–39, 188,
  189; in personal life, 12, 198–99;
  revivalism of (*see* Islamic revivalism and
  activism); and secularism (*see*
  Secularism); threatened by the West,
  51–53, 56, 171–72, 175–76, 177; as
  threat to the West, 3, 4–6, 17, 119, 175–
  76, 177, 184, 205–6; and the U.S., 5, 6,
  68, 178, 212; Western view of, 4–5, 25,
  44–46, 47–48, 60, 77, 83, 113–15, 162,
  173–74, 178–84, 190, 193, 198, 200–202,
  212, 223n.1; women in (*see* Women)

Islam and the Logic of Force (Fadlallah),
    149
Islam in Perspective (Bannerman), 169
Islambuli, Khalid, 96, 137
Islamic activism. *See* Fundamentalism;
    Islamic revivalism and activism
Islamic Alliance, 133
Islamic AMAL, 146, 148
Islamic Association, 157, 158. *See also*
    Islamic Tendency Movement
Islamic Call Society (al-Dawa); in Iraq,
    20; in Libya, 80, 86, 87
Islamic Government (Khomeini), 109
Islamic Group. *See* Jamaat-i-Islami
Islamic Jihad, 11, 161, 182–83
Islamic Liberation Organization (ILO); in
    Egypt, 84, 95, 96, 133–34, 136–37; in
    Jordan, 84, 137; in Libya, 84; members
    of, 136–37; and Sadat, 134, 136–37; in
    Tunisia, 84
Islamic reformism. *See* Modernism
Islamic Resistance, The (Fadlallah), 149
Islamic revivalism and activism, 11–12 (*see
    also* Fundamentalism; *individual
    countries*); causes of, 14, 75–76; as a
    challenge to the West, 207–8, 210–12;
    and democracy (*see* Democracy: and
    Islamic activists); in eighteenth and
    nineteenth centuries, 49–50, 55, 121,
    goals of, 23, 212; and government, 11–
    12, 116–18, 206 (*see also individual
    states*); and the Gulf War, 194–95, 196,
    197; ideological worldview of, 19, 50,
    126; and Iranian Revolution, 4, 12, 18,
    19–20, 145, 209; and Israel, 73–74; and
    Judaism, 26; linked to nazism and
    communism, 168, 208; as mainstream,
    23, 199–200, 202, 209, 212; and
    modernization, 9, 10–11, 14–15, 19, 23,
    164; organizations of, 16, 64, 69, 70,
    119–20, 163–67, 172, 173, 178, 202,
    206–8 (*see also individual
    organizations*); and terrorism, 198, 199,
    202, 206, 208, 209, 210, 211; as threat,
    197–98, 200, 208, 209, 210, 211;
    underestimation of, 197–98, 199, 203–4;
    and the U.S., 168, 169, 192, 207–8, 212;
    West rejected by, 14, 15–17, 69, 171,
    178–79, 192; Western view of, 10, 129,
    168, 169, 170–71, 172–74, 182, 192,
    197–98, 199–200, 202–3, 207–8, 212,
    223n.1 (*see also* Islam: Western view of)
Islamic Salvation Front (FIS, Algeria),
    164, 167; and government, 152, 162,
    166–71, 187, 189; and the Gulf War, 195

Islamic Society. *See* Jamaat-i-Islami
Islamic Tendency Movement (MTI,
    Tunisia), 152, 153, 154, 167 (*see also*
    Islamic Association; Renaissance Party);
    and Bourguiba, 158–59, 160–62
Israel, 18 (*see also* Palestine); and Britain,
    19, 120; as colony of Europe and the
    U.S., 72, 73, 109; creation of, 4, 71, 72,
    73, 75, 120, 171; and Egypt, 13, 17, 72,
    73, 94, 98, 120, 139, 170 (*see also* Arab-
    Israeli war [1967]; Camp David
    Accords); and France, 19; and Gulf
    War, 194; and Iran, 22, 109; and Iraq,
    196; and Lebanon, 140, 141, 144, 145,
    146, 147, 171, 196; opposition to, 72, 73,
    85, 125, 132, 146, 171, 182–83; occupied
    territories of (*see* Gaza Strip; Golan
    Heights; Jerusalem; Sinai; West Bank);
    and the U.S., 13, 16, 17, 19, 109, 120,
    125, 127, 146, 178, 194, 196, 207, 208;
    Western support of, 19, 71, 174
Istanbul, 42, 43. *See also* Constantinople
Istiqlal party (Morocco), 65
Italy, 32, 40, 43, 45, 171; and Libya, 80

Jalal-e-Ahmad, 105, 106, 107
Jamaah al-Islamiyya. *See* Islamic
    Association
Jamaat al-Jihad (Holy War), 84; and
    assassination of Sadat, 86, 99, 134, 135;
    in Egypt, 20, 96, 99, 131, 133, 134–35,
    138; and terrorism, 4, 11, 119, 133,
    206–7
Jamaat al-Muslimin. *See* Society of
    Muslims
Jamaat-i-Islami (Pakistan), 67, 69–70, 108,
    120, 124, 154 (*see also* Mawdudi,
    Mawlana Abul Ala); in Afghanistan,
    130; in Bangladesh, 130, 187; Egypt, 94,
    133; and the Gulf War, 196, 197; in
    India, 120, 130, 187; in Kashmir, 130,
    187; media coverage of, 202;
    membership of, 136; in politics, 129–30,
    187; and the West, 64, 69, 124
Jamiyyat al-Islah (Kuwait), 187, 202
Jerusalem: and the Crusades, 40–42, 170,
    175; Israeli capture and occupation of,
    12–13, 133
Jihad (concept), 177, 178, 190, 194, 202;
    and European colonialism, 170; and
    Islamic revivalism, 19, 50, 73–74, 128,
    129, 135, 150, 165; and Khomeini, 109;
    and the Prophet, 53; and Palestine, 73–
    74, 150; in the Quran, 28, 32–33;
    Saddam Hussein's call for, 193, 195

Jihad (Lebanon), 146, 148
Jinnah, Muhammad Ali, 66, 67
Jordan, 11, 13, 63, 68, 71; and democracy, 185, 187; and the Gulf War, 194, 195, 196; Islamic revivalism in, 4, 11, 23, 70, 165, 185, 187, 198, 206; ulama in, 194
Judaism, 180, 198, 192, 199, 208; in Arabia, 28, 29, 188; and Christianity, 26, 38; and Islam, 26, 30, 31, 33, 39, 81, 135, 170, 188; and Israel, 13, 182, 196; and Jerusalem, 41; and law, 34; origins of, 26; and terrorism, 174, 198
Jund Allah. *See* Army of God

Kajar dynasty, 102
Kashmir, 4, 11, 70, 182, 183
Khadija, 27
Khamenei, Ayatollah, 116, 194
Khan, Sayyid Ahmad. *See* Ahmad Khan, Sir Sayyid
Kharaji, 204
Khomeini, Ayatollah Ruhollah, 22, 71, 106, 108, 140, 203; criticism of, 95, 117, 136; death of, 116, 151, 209; discrediting of, 203; during Shah's rule, 11, 105, 111; government of, 204, 205; governments of Middle East criticized by, 20–21, 114; and the Gulf War, 194; and Iranian Revolution, 18, 20, 79, 108, 114, 147, 170 (*see also* Iranian Revolution: export of); and Islamic AMAL, 146; and Islamic revivalism, 8, 12, 18, 21, 79, 145, 149, 163, 182, 207; and Lebanon's Hizbullah, 142, 146, 147, 148, 149–50, 151; and Rushdie affair, 116, 190–91; and terrorism, 87, 169, 180, 207, 209; and the U.S., 3, 101, 114, 169; and the West, 18, 101, 108–9, 166, 169, 170, 179, 180, 204, 207
Kishk, Abd al-Hamid, 100, 139–40
Knollys, Richard, 42
Krauthammer, Charles, 175, 182–83
Kuwait, 15, 22, 148, 184; and democracy, 185, 187; Iraqi invasion of (*see* Gulf War); Islamic revivalism in, 23, 206

Labor party (Egypt), 133
Lebanese civil war (1975–76), 12, 141, 144, 149, 151, 197
Lebanese Resistance Battalions. *See* AMAL
Lebanon, 13, 71, 141, 171, 183; and Britain, 19, 141; Christianity in, 14, 16, 140, 141, 142, 143, 144, 146, 181; civil war in (*see* Lebanese civil war); Druze

of, 142, 144; and France, 19, 51, 63, 68, 141, 148, 150; and the Gulf War, 196, 197; hostages taken in, 116, 119, 140; and Iran, 22, 116, 141, 142, 146–47, 148, 149–50; and Iranian Revolution, 22, 114, 144, 145; and Iraq, 142, 147; Islam in, 11, 182; Islamic revivalism in, 10, 140–41, 147, 151, 164, 203; and Israel, 140, 141, 144, 145, 146, 147, 171, 196; Maronites of (*see* Lebanon: Christianity in); modernization in, 23, 160, 164; and Palestine, 140, 141, 149, 150; religion in, 141–42; and Saudi Arabia, 151; Shii of, 14, 140, 142–45, 146–47, 149, 150, 151; Sunni of, 142, 143, 144, 149; and Syria, 141, 151; and terrorism, 5, 119; and the U.S., 16, 19, 146, 148, 150–51, 178, 192, 207; and the West, 19, 140, 141, 150, 151, 166, 197, 209
Lewis, Bernard, 173–74, 177–79
Liberation Movement of Iran, 106
Libya, 80, 81, 82, 85; and Africa, 86, 87; Christianity in, 80; and Egypt, 79, 85, 86, 117; foreign policy of, 85–87; and the Gulf War, 196; Islam in, 12, 79, 80–81, 85, 89, 116 (*see also* Qaddafi: and Islam); and Islamic revivalism, 8, 11, 50, 83, 86, 87, 198, 203, 208 (*see also* Qaddafi: and Islam); nationalism in, 70, 78–79, 80, 182; oil in, 80, 85, 86, 155; opposition movements supported by, 86–87; revolution in, 70, 83, 85; Revolutionary Command Council in, 81, 86; and Saudi Arabia, 85, 87; and the Soviet Union, 85, 87; and the Sudan, 79, 91, 92, 117; and terrorism, 8, 85, 86, 208; and the U.S., 11, 80, 85, 87, 117, 184, 208; and the West, 67, 87, 184, 208

Maalim fil Tariq (Qutb), 128
Madani, Abbasi, 154, 164, 167, 195; and democracy, 166, 189
Maghreb (*see also* Algeria; Morocco; North Africa; Tunisia): Islamic revivalism in, 151–53; Sufis of, 152; ulama in, 152
Mahathir, Muhammad, 12
Mahdi (Sudan), 50, 88, 90
al-Mahdi, Sadiq, 90, 91, 92, 205
Mahfouz, Naguib, 191
Mahmud, Shayk Abdul Halim, 136
Malaysia, 13, 20, 62, 187; British colonization of, 51, 68; Chinese-Malay

Malaysia (*continued*)
riots in, 12, 14; and the Gulf War, 195,
196; Islamic revivalism in, 11, 12, 23,
203, 206; and the U.S., 195–96
Malaysian Youth Movement (ABIM), 20
Mao Tse-tung, 81
Maoists, 86
Martel, Charles, 175
Marxism, 70, 73, 81, 122, 123, 191 (*see
also* Communism); rejected by Muslims,
15, 59, 69, 85; in Tunisia, 156, 160
Mauretania, 196, 197
Mawdudi, Mawlana Abul Ala, 67, 120–23,
124–25, 126, 128 (*see also* Jamaat-i-
Islami); criticism of West by, 121–22,
129; influence of, 120, 130, 134, 152,
154, 155; and Islam, 70, 108; and
Pakistani politics, 130
Mecca, 35, 83, 162, 184, 194; and
Muhammad, 28–29, 30, 53, 136; Saudi
Arabia's claims on, 17, 74
Medina, 29, 31, 35, 43, 194; and
Muhammad, 29–30, 136; Saudi Arabia's
claims on, 17, 74
Mehmet the Conqueror, 43
Middle East (*see individual countries*):
Islam in, 12, 13, 17, 20, 78–79, 175, 176
(*see also individual countries*); Libya
condemned by, 87; scholars of, 201–2,
203–4
Milestones (Qutb), 128
*Miseries and Tribulations of the Christians
Held in Tribute and Slavery by the Turks*
(Gregevich), 43
MNLF. *See* Moro National Liberation
Front
Modernism, 55–62, 64, 69, 121, 201;
criticism of, 124–25, 179; and
nationalism, 59–60, 62, 63
Modernization, 36 (*see also individual
countries*); and Christianity, 48; and
Europe, 8–9; impact of, 16, 51; and
Islam, 9, 13, 16–17; and Islamic
revivalism, 9, 10–11, 14–15, 23, 164;
rejection by Muslim countries of, 14,
15–17; as response to colonization, 53–
54; and secularization, 9, 11, 15–16, 54–
55, 68, 200; and Western world, 8–9,
15–16, 67–68; and Westernization, 200,
202
Mogul, 42, 50
Moro National Liberation Front (MNLF),
86–87
Morocco, 55, 152, 154, 183 (*see also*
Maghreb; North Africa); and the Gulf

War, 195, 196, 197; Islamic revivalism
in, 4, 11, 152, 166, 171, 181, 206;
nationalism in, 62, 64–65
Mossadegh, Muhammad, 103––4, 106
Mourou, Abdelfattah, 156, 163, 167
Movement for the Dispossessed, 14, 143,
144. *See also* AMAL
MTI. *See* Islamic Tendency Movement
Mubarak, Hosni, 94, 131; criticism of, 98–
99; and Islam, 97–100; and Islamic
revivalism, 132–33, 135
Muhammad ibn Abdullah, Prophet, 6,
145, 188; birth of, 27; Christian view of,
25, 38, 45, 46; death of, 30, 31, 34;
establishment of community-state in
Medina by, 17, 18, 29–30, 33, 53, 136;
example of, 7, 19, 57, 82, 120, 122, 123;
and Islamic revivalism and reform, 50;
message of, 26–28, 29, 30, 32, 34, 35
(*see also* Quran); and monogamy, 58;
and Qaddafi, 8, 81, 82, 83; in *The
Satanic Verses*, 191
Muhammad, Prophet. *See* Muhammad ibn
Abdullah, Prophet
Muhammad's youth. *See* Islamic
Liberation Organization
Mujahidin, 20
al-Muswi, Abbas, 148
Musawi, Hussein, 146
Muslim Brotherhood, 108 (*see also* al-
Banna, Hassan); as alternative to
capitalism and Marxism, 122, 123; in
Egypt, 11, 20, 64, 69–70, 74, 84, 94, 95,
98, 99–100, 120, 126, 129, 130, 131,
132–33, 136, 137, 140, 154, 155, 157,
159, 164, 166, 187, 195, 197, 202;
factions of, 130, 131; in the Gulf
states, 132; and the Gulf War, 194, 195,
197; in Iraq, 132; and Israel, 132;
in Jordan, 11, 187, 195; membership of,
123, 136; moderate reformism of, 131–
33; and Mubarak, 131, 132–33; and
Nasser, 128, 129, 130, 133–34, 137; and
Palestine, 125; and politics, 128–29, 131,
140, 187; and Qaddafi, 8, 84; and Qutb,
126, 127, 128, 129; and Sadat, 131–32,
133, 137; in Saudi Arabia, 132; in the
Sudan, 11, 84, 88–89, 91, 93, 157, 187;
suppression of, 129, 130, 131, 137; in
Syria, 74; and terrorism, 132; and
Tunisia, 154, 155, 157, 159; and the
U.S., 132; West criticized by, 120, 124,
125
Muslim Family Reforms, 132
Muslim League, 60, 66–67

Mustafa, Shukri, 136, 137
Mutiny of 1857, 58

Nadwi, Abul Hasan Ali, 67
Nasir al-Din Shah, 102
Nasser, Gamal Abdel, 85, 95, 100, 139;
anti-imperialist views of, 71, 72; and
Arab nationalism/socialism, 17, 69, 70,
79, 104 (*see also* Arab nationalism); and
Arab-Israeli war (1967), 17, 76;
influence of, 69, 70, 71–72, 87, 94, 154;
and Islam, 74–75, 94, 97; and Muslim
Brotherhood, 128, 129, 130, 133–34,
137; and Saudi Arabia, 71, 74–75
Nasserism, 70, 72, 104 (*see also* Arab
nationalism)
National Front, 88, 106
National Islamic Front (NIF), 93
National Salvation Committee, 146
Nationalism, 13, 58 (*see also individual
countries*); Arab (*see* Arab nationalism;
Nasserism); fathers of, 62; and Islam,
11, 62–63, 67, 78; and Islamic
modernism, 59; liberal, 15, 47, 68, 70,
71, 72, 76, 120, 183; local transformed
into transnational, 72; and Pan-Islam,
60; as a reaction to Western
imperialism, 62, 66; and secularization,
15, 59, 62, 64, 67, 69, 73, 206, 210; and
umma, 62
Nazism, 168, 208
Neglected Obligation, The (al-Farag), 96–
97, 134
Nehru, 72
Neorevivalism, 124–26
Netherlands, 51
New Woman, The (Amin), 58
New World Order, 4, 5–6, 169, 196, 211
Nidal, Abu, 86
NIF. *See* National Islamic Front
Nigeria, 13, 50, 188, 191, 196
Nimeiri, Gaafar Muhammad, 70, 87, 162;
government of, 79, 89, 204; and Islam,
12, 79, 88, 89–92, 163; and Qaddafi, 87,
117, 184; and the U.S., 89, 90, 91–92,
178, 207
North Africa, 38, 43, 176 (*see also* Africa;
Algeria; Maghreb; Morocco; Tunisia);
and democracy, 189, 210, 211; and
France, 51, 64, 65, 68; Islamic
modernism in, 61, 64; Islamic revivalism
in, 152, 166, 171, 203; nationalism in,
63, 64
Norton, Augustus Richard, 142

OIC. *See* Organization of the Islamic
Conference
Oil, 15, 17, 94. *See also individual states*
Organization for Holy War. *See* Jamaat al-
Jihad
Organization of the Islamic Conference
(OIC), 74, 117, 191, 193
Orientalism, 16, 202
Ottoman empire, 54, 120, 179; and Arab
nationalism, 63; divided up by Allies,
63, 66, 75, 121, 171, 182; and Europe, 3,
42–44, 45, 170; and eighteenth-century
Islamic revivalism, 50; First Ottoman
Constitution of, 54; support of Germany
during World War I, 63, 66, 121; and
ulama, 44

Padri, 50
Pahlavi, Muhammad Reza Shah. *See* Shah
of Iran
Pahlavi, Reza Shah, 102, 106, 171
Pahlavi dynasty, 101, 104, 112
Pakistan, 14, 68, 188, 189 (*see also* South
Asia); civil war in (*see* Pakistan-
Bangladesh civil war); East (*see*
Bangladesh); government in, 185, 204;
and the Gulf War, 170, 195, 196, 197;
Islam in, 4, 13, 89, 117; Islamic
revivalism in, 8, 11, 12, 14, 23, 64, 79,
154, 183, 185, 187–88, 198, 203, 204,
206; nationalism in, 62, 67; and Rushdie
affair, 190, 192; Shii of, 20; Sunni of, 20;
and the U.S., 8, 11, 117, 184, 192, 196;
and the West, 204
Pakistan-Bangladesh civil war (1971), 12, 14
Palestine, 120, 171, 174 (*see also* Israel);
British colonization of, 51, 63; and
Egypt, 71, 72; and France, 63; and the
Gulf War, 194, 197; Intifada in (*see*
Intifada); and Iraq, 197; and Lebanon,
140, 141, 149, 150; liberation of, 13, 71,
74, 149, 150; occupation of, 19, 125,
139, 182 (*see also* Gaza Strip; Golan
Heights; Jerusalem; Sinai; West Bank)
Palestine Liberation Organization, 98, 144
Party of God. *See* Hizbullah
PAS (Malaysia), 20, 187
Persian empire, 31, 38, 39, 101, 104
Peters, Francis, 39
Pfaff, William, 3
Philippines, 11, 86–87, 115
Pipes, Daniel, 223n.1
Piscatori, James, 179–80, 184
P.L.O. *See* Palestine Liberation
Organization

Poland, 208
Prophet. *See* Muhammad ibn Abdullah, Prophet
Protestant Reformation, 44, 46, 56, 59
Protestantism, 7

Qaddafi, Muammar, 144, 145, 151, 184; coup d'état of, 79–80, 87; foreign policy of, 85–87; and Islam, 8, 12, 79, 80, 83–85, 117, 163, 184, 208; Muslim reaction to, 83, 84–85; revolution of, 70, 79–80, 81–82, 83; socialism of, 80, 81–82, 83, 204; and terrorism, 8, 79, 85, 169, 208, 209; Third International Theory of, 81, 83; and the West, 81, 169, 170, 204, 208
Quayle, Dan, 168, 192, 208
Quran, 25, 78, 109; and Islamic revivalism and reform, 50, 120; jihad in, 28, 32–33 (*see also* Jihad); message of, 7, 17, 19, 26–28, 29, 30, 33, 34, 35, 58, 81, 122, 125–26, 149, 188 (*see also* Muhammad ibn Abdullah, Propet; message of); and Qaddafi, 82, 83; in *The Satanic Verses,* 191
Quran Preservation Society (Tunisia), 155, 156, 157
Quraysh, 29
Qutb, Muhammad, 155
Qutb, Sayyid, 131, 134, 137; influence of, 70, 120, 126, 130, 152, 154, 155; West criticized by, 127–28, 129, 135

Rafsanjani, Hashemi, 116, 151, 197
al-Rahman, Shayk Umar Abd, 137
Reagan, Ronald, 87, 113, 208
Reconstruction of Religious Thought in Islam (Iqbal), 59
Red Book (Mao Tse-tung), 81
Reform Society. *See* Jamiyyat al-Islah
Renaissance Party (Tunisia), 154, 166, 167, 202 (*see also* Islamic Tendency Movement); and politics, 152, 162–63, 187
Renan, Ernest, 46
Republican Brothers (Sudan), 91
Resurrection party. *See* Baath party
Revolutionary Guards (Iran), 22
Revolutionary Justice Organization, 148
Rida, Rashid, 64, 120
Rodinson, Maxime, 37, 45
Roman empire, 3, 31, 40
Romania, 176
"Roots of Muslim Rage" (Lewis), 173–74, 177–79
Rushdie, Salman, 5, 176, 180, 188,

190–93; opposition to, 3, 116, 170, 190–91, 192

Sasanid empire, 31
Sadat, Anwar, 160, 162, 184; assassination of, 93, 96, 97, 99, 119, 131, 132, 134, 135, 137, 209; criticism of, 94–96, 97, 100, 139; and Islam, 12, 17, 79, 94–97, 117, 131, 133; Khomeini criticized by, 95, 117; and Muslim Brotherhood, 131–32, 133, 137; Shah of Iran supported by, 94, 95, 97, 117; and the Sudan, 87–88; and the West, 94, 95, 96, 97, 101, 139
al-Sadr, Muhammad Baqir, 142, 147
Sadr, Musa, 14, 142–45, 147, 149, 151
Safavid dynasty, 42, 50, 102
Said, Edward, 202
Salaama, Shaykh Hafiz, 98
Saladin, 42
Salafiyya movement, 57, 64
Salah-al-Din. *See* Saladin
Salvation from Hell, 131, 133; and terrorism, 4, 119, 135, 206
al-Sanusi, Muhammad ibn, 50, 80
Saracens, 40
Satanic Verses, The (Rushdie), 3, 116, 190–93
Saud, house of, 71, 78, 184
Saudi Arabia, 27, 121, 132; government in, 185, 204; and the Gulf War, 194, 195; and Iran, 22, 184; Islam in, 78, 85, 86, 116; Islamic revivalism in, 8, 11, 17–18, 23, 50; and Lebanon, 151; and Libya, 85, 87; and Nasser, 71, 74–75; oil in, 15, 17–18, 74, 155; Shii in, 20, 21; and the Sudan, 90, 91; Sunni in, 20, 21; ulama in, 8, 78, and the U.S., 11, 21, 117, 184, 194–95; and the West, 148, 194–95, 204
Secularism, 78; and Arab nationalism, 15, 69, 74, 75, 78–79; and Baath party, 75, 78; in Egypt, 9–10, 23, 61, 78–79; and Islamic revivalism, 179, 199; and modernization, 9, 11, 15–16, 54–55, 68, 78–79, 200–203, 209; and nationalism, 59, 62, 64, 67, 69, 73, 210; in Turkey, 78; Western, 177, 199, 200–202, 204
Shaarawi, Huda, 58
al-Shaarawi, Shaykh Muhammad Mitwali, 100, 139–40
Shabab Muhammad. *See* Islamic Liberation Organization
Shah of Iran, 11, 56, 160; government of, 54, 56, 65, 102–5; and Islam, 16, 18, 79,

102–3, 104; opposition to, 11, 18, 105–6, 108, 109, 110, 111, 112, 115; overthrow of, 18, 102, 103, 111, 112, 140; rule of, 102–4; Sadat's support for, 94, 95, 97, 117; and the U.S., 16, 18, 68, 104, 178, 207 (*see also* Iran: and the U.S.); and the West, 18, 101, 104, 171 (*see also* Iran: and the West)

Sharia: in Egypt, 96, 135; and European colonialism, 48–49; in Iran, 109, 115; and Islamic revivalism, 19, 50, 118, 122, 126, 164; and Khomeini, 112; in Libya, 83; and modernization, 36, 68; and Muhammad, 28, 35; and Saudi Arabia, 78; in the Sudan, 89, 90, 91–92, 93; in Tunisia, 153

Shariati, Ali, 112; and Iranian reform, 105, 106–8, 110, 111, 205

Shariatmadari, Ayatollah, 113

Shii Front for the Liberation of Bahrain, 22

Shii Islam, 204 (*see also individual countries*); and Iranian Revolution, 18, 20; origins of, 30–31

al-Shirazi, Ayatollah Hasan, 65

Shukry, Ibrahim, 38, 40, 98

Siddiqui, Kalim, 176

Signposts (Qutb), 128

Sinai, 12, 72

Siriya, Salih, 137

Six-Day War. *See* Arab-Israeli war (1967)

Social Basis of the Third International Theory (Qaddafi), 81

Social Justice in Islam (Qutb), 127

Socialist Labor Party (Egypt), 98

Society of Muslims. *See* Takfir wal-Hijra

Solution of the Economic Problem: Socialism (Qaddafi), 81

Solution to the Problem of Democracy (Qaddafi), 81

South Africa, 196. *See also* Africa

South Asia, 63, 66–67, 68, 211. *See also* India

Southeast Asia, 87, 195 (*see also* Asia; South Asia); Islam in, 4, 13, 20, 37, 49, 171; political liberalization in, 210, 211

Southern, R. W., 45

Soviet Union, 32, 51, 183, 196; breakup of, 4, 174, 175 (*see also* Communism: collapse of); and Egypt, 72, 97; and Iran, 22, 51, 65, 103, 108, 171, 191; and Islam, 4, 175, 184, 191; and Libya, 85, 87; and Syria, 151; and terrorism, 87; and the U.S., 3–4, 76, 168, 169 (*see also* Cold War)

Spain, 3, 32, 45, 171; and early Islamic expansionism, 38, 40

SPLA. *See* Sudan People's Liberation Army

SPLM. *See* Sudan People's Liberation Movement

Sudan, 181; and Britain, 90; Christianity in, 88, 188; and democracy, 187; and Egypt, 79, 87, 91; and the Gulf War, 170, 196, 197; and Iran, 92; and Iranian Revolution, 115; Islam in, 12, 23, 79, 87–93, 117; Islamic revivalism in, 11, 50, 70, 88, 91–92, 187–88, 198, 206; and Israel, 13; and Libya, 79, 91, 92, 117; nationalism in, 70, 87; and Saudi Arabia, 90, 91; Sufis in, 91; and the U.S., 87, 89, 90, 91–92, 117, 178, 207; and the West, 68, 115, 204

Sudan People's Liberation Army (SPLA), 90

Sudan People's Liberation Movement (SPLM), 90, 93

Suez Canal Company, 72

Suez crisis (1951), 71, 72

Sufism, 12, 36–37, 49, 65; in Egypt, 94, 100, 132, 139; of the Maghreb, 152; in the Sudan, 91

Sukarno, 72

Suleiman the Magnificent, 43

Sunnah, 7, 50, 122

Sunni Islam, 18, 20, 21, 204. *See also individual countries*

Supreme Islamic Shii Council, 143

Syria, 13, 15, 38, 80, 197; and early Islamic expansionism, 38; and Egypt, 73; French colonization of, 51, 63; and Iran, 147, 151; and Iraq, 183; Islam in, 70, 183; and Lebanon, 141, 151; nationalism in, 63, 70, 71, 78–79; and the Soviet Union, 151; and terrorism, 87, 151; and the West, 151, 171

Tabligh-i Jamaat, 202

Taha, Mahmud Muhammad, 91

Taif Accords, 151

Takfir wal-Hijra (Egypt), 11, 133, 134, 136, 137

Terrorism, 177, 180 (*see also individual countries and organizations*); associated with Islam, 4, 5, 23, 25, 77, 83, 114, 164, 170, 173, 178, 198, 199, 202, 206–7, 208, 209, 210, 211

Thailand, 11, 86

al-Thalabi, Abd al-Aziz, 55, 65

Third World, 72, 85, 107, 108, 156; revolution in, 87, 114; and the U.S., 92, 196; worldview of, 71

Tilmassani, Omar, 131, 132

Tobacco Protest (Iran, 1891–92), 56, 65, 102, 103, 110

Transjordan, 51, 171

Treaty of Sèvres, 63

Tripoli Agreement, 87

Tripoli Pact, 79

al-Tufayli, Subhi, 148

Tunisia, 79, 98, 185 (*see also* Maghreb; North Africa); and France, 68, 153, 154, 155, 158, 161; government in, 79, 184, 185, 187, 211; and the Gulf War, 170, 195, 196, 197; and Iran, 159, 160, 161; and Iranian Revolution, 115, 156, 157; Islamic modernism in, 55; Islamic revivalism in, 10, 11, 12, 23, 130, 152, 153, 154–63, 164, 165, 166, 169, 171, 187, 198, 203, 206, 208; modernization in, 23, 153, 164; nationalism in, 62, 154; ulama in, 153; and the U.S., 153; and the West, 115, 152, 153, 154, 155–56, 158, 159

al-Turabi, Hassan, 88–89, 91, 92–93, 154

Turkey, 51, 66, 108, 176; and Islam, 12, 206; as secular state, 78, 153

Twelfth Imam, 110

Uganda, 86

UGTT. *See* General Union of Tunisian Workers

Ulama, 37, 57, 201 (*see also individual countries*); and Islamic revivalism and reform, 50, 124, 134–35, 136; and jihad, 134–35; and modernism, 56, 58–59; and modernization, 54–55; and Qaddafi, 83, 84

Umayyad Caliphate, 31, 110

Ummah: and European colonialism, 48–49, 51, 55–56, 57; and Islamic modernism, 55–56, 57; and Islamic revivalism, 50, 120, 165; and Muhammad, 17, 19, 29–31; and nationalism, 62, 64; non-Muslims under, 38–39; and Palestine, 13; and Sadat, 97

United Arab Emirates, 15

United Arab Republic, 73

United Nations, 72, 150–51; and the Gulf War, 195–96, 197

United States, 42, 70, 104 (*see also similar subheadings under* West); and Bahrain, 184; and Europe, 169, 196, 208; and Egypt, 19, 72, 95, 98, 117, 139, 184; and

the Gulf states, 21, 114; and the Gulf War, 194–95; and Indonesia, 195; and Iran, 11, 16, 18, 19, 22, 68, 101, 104, 108, 116, 117, 171, 178, 182, 184, 191, 192, 197, 207; and Iraq (*see* Gulf War); and Islamic governments, 169, 207, 208–9; Islamic revivalism viewed in, 3, 5, 168, 169, 192, 207–8, 212; and Israel, 13, 16, 17, 19, 109, 120, 125, 127, 146, 178, 194, 196, 207, 208; and Jordan, 68; and Kuwait, 184 (*see also* Gulf War); and Lebanon, 16, 19, 146, 148, 150–51, 178, 192, 207; and Libya, 11, 80, 85, 87, 117, 184, 208; and Malaysia, 195–96; Muslims in, 4, 13, 175; neoimperialism of, 75, 85; and Pakistan, 8, 11, 117, 184, 192, 196; policies toward Islam, 5, 6, 16, 91–92; and Rushdie affair, 190; and Saudi Arabia, 11, 21, 117, 184, 194–95 (*see also* Gulf War); and Soviet Union, 3–4, 76, 168, 169 (*see also* Cold War); and the Sudan, 87, 89, 90, 91–92, 117, 178, 207; and the Third World, 196; and Tunisia, 153; viewed by Muslims, 68, 69, 75, 91–92, 127–28, 132, 146, 148, 178, 207

Urabi nationalist revolt, 56, 57

Urban II (Pope), 41

Vance, Cyrus, 182

Vanderwalle, Dirk, 161

Voice of the Islamic Revolution (Iran), 114

Voll, John, 51, 190

Wafd party (Egypt), 133

Wahhabi, 50, 121, 204

West (*see also similar subheadings under* Britain; Europe; France; United States); and boundaries of Arab world, 72, 75; Crusades mythologized in, 39–40, 42, and democracy in Middle East, 184–86, 208; and Egypt, 19, 68–69, 71–72, 94, 95, 97, 98, 99, 139; impending confrontation with Islam of, 175–76, 177; imperialism of, 46, 51, 52, 62, 69, 72, 74, 75, 107, 108, 125, 178, 193 (*see also* European colonization); interests in Muslim countries of, 4, 17, 194; and Iran, 8, 16, 18, 19, 22, 101, 102, 103–4, 105, 106, 107, 108–9, 111, 116, 140, 184, 191–92, 197, 198, 204, 207; and Iraq, 115 (*see also* Gulf War); Islam as threat to, 3, 4–6, 10–11, 17, 47–48, 92; Islam in, 20, 175–77, 191; Islam viewed in (*see*

Islam: Western view of); Islamic dependence on, 81, 186, 194, 206; and Islamic modernism, 55–62; and Islamic organizations, 163, 164, 165–66, 167 (*see also individual organizations*); Islamic revivalism viewed in (*see* Islamic revivalism and activism: Western view of); and Israel, 19, 71, 174; Judaeo-Christian heritage of, 135, 170, 177, 179, 186, 205; and Lebanon, 19, 140, 141, 150, 151, 166, 197, 209; and Libya, 68, 87, 184, 204, 208; and modernization of Islamic states, 8–9, 13, 15–16, 51, 67–68, 77–78; and Morocco, 152; Nasser's challenge to, 71–72; and Ottoman empire, 170; and Pakistan, 204; relations with Islamic world, prior to 11th century, 34, 40; religion in (*see* Secularism: Western); and Rushdie affair, 190–93; and Saudi Arabia, 148, 194–95, 204 (*see also* Gulf War); and the Sudan, 68, 115, 204; and Syria, 151; as threat to Islam, 51–53, 56, 92, 128–29; and Tunisia, 115, 152, 153, 154, 155–56, 158, 159; viewed by Muslims, 14–15, 51, 52–53, 55, 68, 69, 75–76, 135, 150, 171–72, 177–78, 186, 190, 205
West Bank: Islam in, 4, 11, 182; Israeli capture and occupation of, 12, 13, 171, 182–83, 195–96; and the U.S., 207 (*see also* Israel: and the U.S.)
White Revolution, 8, 101, 103
Why the Islamic Way? (Nimeiri), 88
Women: and Islam, 5, 6, 48; and Islamic revivalism, 84, 138, 167, 188; and modernism, 58, 122; and nationalism, 58
World Bank, 88, 91
World Muslim League, 74
World War I, 45, 51, 63, 64, 65, 68; effect on Islamic world of, 181–82; Ottoman and German alignment during, 66, 121
World War II, 45, 68, 75

Yassin, Abdesallem, 154, 166
Yazid, 110, 143
Yemen, 196
Young Turks, 63
Yugoslavia, 4, 181, 182

Zaghlul, Saad, 61, 64
Zakaria, Fouad, 170
Zionism: and criticism of the West, 16, 19, 125, 135, 146; and Islamic revivalism, 122, 125, 134, 146
al-Zumur, Abbud, 137